EQUITY AND THE ENVIRONMENT

RESEARCH IN SOCIAL PROBLEMS AND PUBLIC POLICY

Series Editors: William R. Freudenburg and
Ted I. K. Youn

Recent Volumes:

RESEARCH IN SOCIAL PROBLEMS AND
PUBLIC POLICY   VOLUME 15

EQUITY AND THE ENVIRONMENT

EDITED BY

ROBERT C. WILKINSON

*Bren School of Environmental Science and Management,
University of California, Santa Barbara, CA, USA*

WILLIAM R. FREUDENBURG

*Environment and Society Environmental Studies Program,
University of California, Santa Barbara, CA, USA*

ELSEVIER
JAI

Amsterdam – Boston – Heidelberg – London – New York – Oxford
Paris – San Diego – San Francisco – Singapore – Sydney – Tokyo
JAI Press is an imprint of Elsevier

JAI Press is an imprint of Elsevier
Linacre House, Jordan Hill, Oxford OX2 8DP, UK
Radarweg 29, PO Box 211, 1000 AE Amsterdam, The Netherlands
525 B Street, Suite 1900, San Diego, CA 92101-4495, USA

First edition 2008

British Library Cataloguing in Publication Data
A catalogue record for this book is available from the British Library

ISBN: 978-0-7623-1417-1
ISSN: 0196-1152 (Series)

For information on all JAI Press publications
visit our website at books.elsevier.com

Printed and bound in the United Kingdom

08 09 10 11 12 10 9 8 7 6 5 4 3 2 1

CONTENTS

PART III: EQUITY AND INEQUALITY IN EXTREME ENVIRONMENTS

PART IV: INEQUALITY AND EQUITY AS PREDICTORS OF ENVIRONMENTAL HARM

LIST OF CONTRIBUTORS

Lisa M. Berry	Donald Bren School of Environmental Science and Management, University of California, Santa Barbara, CA, USA
James K. Boyce	Department of Economics, University of Massachusetts, Amherst, MA, USA
Anthony J. Brazel	School of Geographical Sciences, Arizona State University, Tempe, AZ, USA
Robert D. Bullard	Environmental Justice Resource Center, Clark Atlanta University, Atlanta, GA, USA
Christine Evans	University of Wisconsin-Parkside, Kenosha, WI, USA
Bruce C. Forbes	Arctic Centre, University of Lapland, Rovaniemi, Finland
William R. Freudenburg	University of California, Santa Barbara, CA, USA
Sharon L. Harlan	School of Human Evolution and Social Change, Arizona State University, Tempe, AZ, USA
G. Darrel Jenerette	Department of Botany and Plant Sciences, University of California, Riverside, CA, USA
Nancy S. Jones	Global Institute of Sustainability, Arizona State University, Tempe, AZ, USA
Larissa Larsen	Urban and Regional Planning Program, University of Michigan, Ann Arbor, MI, USA

Paul Mohai	School of Natural Resources & Environment, University of Michigan, Ann Arbor, MI, USA
Lela Prashad	School of Earth and Space Exploration, Arizona State University, Tempe, AZ, USA
Anne Statham	Department of Sociology, University of Wisconsin-Parkside, Kenosha, WI, USA
William L. Stefanov	Image Science & Analysis Laboratory, NASA Johnson Space Center, TX, USA
Dorceta E. Taylor	School of Natural Resources & Environment, University of Michigan, Ann Arbor, MI, USA
Robert C. Wilkinson	Bren School of Environmental Science and Management, University of California, Santa Barbara, CA, USA

EQUITY AND THE ENVIRONMENT: A PRESSING NEED AND A NEW STEP FORWARD

William R. Freudenburg and Robert C. Wilkinson

Within months of the time when the famous Santa Barbara oil spill of 1969 helped to inspire the first "Earth Day," the academic world joined in a virtual explosion of societal interest in a topic that inherently lies in the confluence between "social problems" and "public policy" – the ways in which humans use and abuse the natural environment. In the worlds of social movements and public policy, that newfound interest showed up in dramatic growth of environmental organizations and the passing of powerful new environmental laws. In the academic world, echoes of the explosion showed up in equally dramatic growth of interdisciplinary "environmental" programs. Equally importantly, many of the new programs included something that was seen as radical on many campuses at the time, namely an explicit focus on the fact that "environmental problems" are inherently social problems as well.

In some senses, the very vigor of the field's growth can be taken as an indication that this seemingly radical approach was long overdue. Up through most of the 19th century, after all, "environmental problems" generally meant the things that "nature" did to people – fires, floods, famines, plagues, droughts, and more. By the time the 20th century moved past its midpoint, however, it had become clear that almost all of the problems we call "environmental" – from holes in the ozone above us to

Equity and the Environment
Research in Social Problems and Public Policy, Volume 15, 1–18
Copyright © 2008 by Elsevier Ltd.
ISSN: 0196-1152/doi:10.1016/S0196-1152(07)15011-0

contaminants in the groundwater below – involve the things that people do to nature. As David Orr (1992) noted, in one of the most sharply distilled observations about the interdisciplinary complexity of environmental issues ever offered, "the symptoms of environmental deterioration are in the domain of the natural sciences, but the causes lie in the realm of the social sciences and humanities."

In many ways, the spirit of Earth-Day innovation has continued to expand. By the time of the first extensive survey of interdisciplinary "environmental" programs to be carried out in the 21st century, Romero and Silveri (2006) would report the existence of well over a thousand such programs and departments in higher-education institutions in the U.S., alone. As is clear from the Romero and Silveri review, those programs are characterized by a good deal of diversity. Still, in addition to the intended similarity of the focus on environmental problems and solutions, academic offerings on environmental issues are also characterized by another form of similarity that appears to be more accidental than intentional. When the programs first started to spread, around the time of that first Earth Day, issues of equity and the environment were usually relegated to isolated classes on environmental ethics. Today, they still are.

To the editors and authors in this volume, that seems wrong. This volume takes note of the fact that, particularly within the past few years, a small but growing body of research has shown that equity issues need to receive greater attention in academia – not just among activists, and not just as topics for environmental ethics courses, but as the focus of careful, academic study. In many ways, in fact, they are at the core of what we call "environmental" problems; they deserve to be at the core of scholarly and scientific work on environmental problems, as well.

In some ways, work on the topic still owes a debt to two volumes that were first published in 1987, making it doubly appropriate that this volume would be sent to press in the middle of the year 2007, the 20th anniversary of those landmark publications. The first of the two, *Toxic Wastes and Race in the United States*, came from the Commission for Racial Justice, United Church of Christ (1987). As is noted in the chapters in this volume by Bullard (2008), Mohai (2008), and Taylor (2008), *Toxic Wastes and Race* had a galvanizing effect on social science research on environmental issues, particularly in the U.S., making it clear that, while we might all be "Passengers on Space Ship Earth," not all of the passengers were breathing the same quality of air. While that volume has also been criticized just as vigorously as almost any work that is truly influential, Mohai's quantitative analysis in this volume, in particular, makes it clear that subsequent examination has largely borne out the validity of its original argument. As noted in *Toxic Wastes and Race at Twenty* (Bullard, Mohai, Saha, & Wright, 2007), race and poverty are even more important in exposure to hazardous waste facilities than previously reported. As noted by Bullard et al. (2007), in fact, people of color now make up the majority (56%) of the persons living within two miles of such facilities in the United States – and they make up more than two-thirds (69%) of the people living in areas where there are clusters of such hazardous waste facilities.

Second, 1987 saw the publication of the work that brought the topic of "sustainability" into mainstream discussions of environmental issues – the World Commission on Environment and Development (1987) report, *Our Common Future* – often remembered today as "Brundtland report," in honor of the Commission's Chair, Gro Harlem Brundtland, the former Prime Minister of Norway, who has also served as the Director General of the World Health Organization. In a Solomon-like balancing act, her Commission argued that, instead of trying to meet the needs of the present "or" the needs of the future, we should seek to encourage the kind of development "that meets the needs of the present without compromising the ability of future generations to meet their own needs."

Despite the impact of those two volumes, however – one focusing on equity and inequality in exposure to environmental harms today, and the other focusing on equity and inequality across time – the majority of the work that examines relationships between humans and the environment has continued to discuss those relationships in terms of aggregates or averages. Still, as is shown by the chapters in this volume, to continue that bad habit into the future would make about as much sense as the old joke about the statistician with his head in the oven and his rear end in the freezer – the one who reported that, at least "on average," he felt just about right.

This volume is intended not just to call attention to the growing body of work on *Equity and the Environment*, but also to encourage more such research in the future. The volume grows out of a series of lectures and symposia on "Equity and the Environment" that were held on the campus of the University of California, Santa Barbara (UCSB) during 2005–2006. Both that series of events and this volume have been made possible in significant part through support provided by the "Critical Issues in America" endowment in the college of Letters and Science at UCSB.

The central theme of this volume is that, far from being the kind of topic that ought to be relegated to a small pigeonhole, issues of equity and inequality deserve to be absolutely central to future work on the connections between humans and the habitat we share with all other life on earth. In particular, emerging research is showing that there are at least five important and interrelated areas of work where much greater attention is warranted. The first three are ones where the need is to build on important work that has been done, having to do with equity and inequality in the distributions of "bads," in patterns of minimal access to necessary "goods," and across time, involving questions of "sustainability." The other two are areas of still-newer work, assessing the degree of inequality in the *creation* of harms – and the evidence that inequality, itself, may actually increase the overall levels of environmental harm being created. We will say more about each of these five interrelated topics before turning to the chapters themselves in a bit more detail.

(1) *Equity and Inequity in Experience of Harm: Environmental Justice.* Work on environmental justice – much of the pressure for which has come from previously marginalized peoples, and much of the academic leadership for which has come from authors in this volume – has been vital in bringing issues of equity (back) to the attention of the scholarly and scientific community. Although equity/justice issues were noted by some work in the 1970s and earlier, it was not until the just-noted report from the Commission for Racial Justice (1987), *Toxic Wastes and Race in the United States,* that Environmental Justice (or EJ) questions came to be the focus of sustained and serious attention in the worlds of science and policy. The EJ movement has involved rural communities exposed to toxics, farm workers exposed to pesticides, factory workers exposed to toxic pollutants, arctic residents exposed to the persistent organic pollutants, and more. The EJ issue has become a focus of international treaties such as *The Basel Convention on the Control of Transboundary Movement of Hazardous Wastes and Their Disposal,* which seeks to regulate and eventually ban the shipment of hazardous waste from wealthier, industrialized nations to impoverished ones.

Environmental justice debates, however, have raised passions not just among activists, but also among normally mild-mannered researchers – perhaps in part because the issue has such serious implications for the ways in which we think (and teach) about our world, and also because research conclusions can lead to radically differing real-world policy implications. Some articles in peer-reviewed journals have argued that there is little empirical basis for EJ concerns, and others conclude that only income, not race, may be a correlate of exposure to harms. Thanks in significant part to work that has been done by authors whose work appears in this volume, however, most experts who are familiar with the empirical findings now see the overall pattern of evidence as providing much stronger support for the opposite conclusion, particularly in the U.S. Although the findings are surprising to many, they are also surprisingly robust: As least within the United States, race is often a "better" predictor of exposure to environmental harm than almost any other variable – with middle-class Blacks in some studies, for example, being exposed to more risks than lower-class Whites, and with members of minority groups, as noted earlier, making up clear majorities of the population living near toxic waste facilities in the U.S.

The environmental justice literature is perhaps the best-developed body of social science research on issues of equity and the environment, and it is represented in this volume by some of the best-known authors in the field – including (alphabetically) Robert Bullard, Paul Mohai, and Dorceta Taylor, all of whom have dealt extensively with issues of environmental exposure and race. It also includes chapters by Sharon Harlan and colleagues (2008) and by Anne Statham and Christine Evans (2008), who examine, respectively, new approaches to the analysis of the ill effects of heat and of the intersection of environment and gender.

(2) *Equity and Inequality in Access to Basic Resources: Health and Development.* The obverse of most environmental justice debates involves not too much exposure to "bads," but too little access to "goods" – the pressing need of many of the world's people to obtain access to minimal levels of needed resources, starting with food and water. The "haves" of today, of course, enjoy standards of living that would have been unimaginable just a century ago – so much so that the need to eat less and lose weight has become a major health concern of present-day Americans. At the same time, unfortunately, many citizens – including too many in America itself – have just the opposite worry, needing to find enough food to eat, and enough clean water to drink. Intriguingly, there is little disagreement about the basic desirability of improving access to needed resources, but there are often stark disagreements about the best ways to do so.

Previous issues of *Research in Social Problems and Public Policy* have included important discussions of this topic, including, for example, the work by Niazi (1999), who noted that some of the key sources of current urban violence in Pakistan are tied directly to inequitable choices about rural access to water in the past. As the work of Niazi and others has begun to spell out, unequal access to environmental resources can actually worsen the overall levels of environmental harm being created – and the worsening problems of environmental impacts can lead to increasingly significant problems of social disruption, in turn. Particularly notable in that connection is the work in this volume by the noted ecological economist,

James Boyce (2008), and by work among the reindeer herders of northern Europe and Siberia by Bruce Forbes (2008).

(3) *Equity and Inequality in Resource Use Across Generations: Sustainability.* Perhaps the equity issue where environmental writings have the longest track record would have to do with intergenerational equity. In many ways, preserving the earth's valuable and vulnerable systems for future generations could be said to be the core of "environmental" concerns, and in the 1960s and 1970s – as well as long before – a number of leading environmental analysts expressed concern about the kind of world we might be leaving for our descendants. Again for this issue, however, the catalyst for a dramatic increase in societal attention came from a "nonacademic" report that came out in the same year that saw the publication of *Toxic Wastes and Race.* As noted above, it was also in 1987 that the UN's World Commission on Environment and Development released the landmark assessment that has come to be known as "the Brundtland report," *Our Common Future,* leading to a dramatic increase in attention to "sustainability."

Virtually all of the authors in this volume deal in one way or another with issues of sustainability, approaching the topic in contexts that range from some of the poorest regions of the world, in the case of work by Boyce and Forbes, to the work of Harlan and her colleagues in analyzing suburban development practices in one of the fastest growing metropolitan regions of the United States, namely Phoenix. The fourth and fifth issues of equity and inequality, by contrast, are ones that have received much less attention to date.

(4) *Equity and Inequality in the Production of Harm: Disproportionality.* As shown by a small but growing body of work, there is also a need to recognize that the new sustainability discourse takes place not in a vacuum, but in a context where a very different viewpoint has long been dominant – the view, in essence, that environmental harm is "just" an inevitable byproduct of prosperity, and that the main responsibility of the present generation is to maximize economic growth, even if that means serious and/or inequitable environmental harm, because "richer is safer," and because the accumulation of wealth is argued to be the best way to make it possible for future generations to deal with whatever challenges may arise.

That established argument is often reinforced, albeit perhaps inadvertently, by the fact that few leading environmental thinkers of the "Earth Day" era – and since – have been social scientists. Instead, as noted in Sharon Harlan's closing reflections in the chapter by Harlan et al. (2008) in this volume, most of the widely recognized "environmental" analysts have tended to be biophysical scientists – experts on other species and physical systems, not on humans. These experts have rarely been comfortable thinking or writing about humans except in terms of global population totals, population density, or other forms of interchangeability.

As spelled out in the chapter in this volume by Lisa Berry (2008), however, a small but growing body of work suggests the common approaches of the past may be badly misleading. This work, sometimes discussed under the heading of "disproportionality," argues instead that there may well be at least as much inequality in the *producing* of harms – even after controlling for economic scale and the specific forms of production involved – as in the *experiencing* of harms. As Berry's chapter spells out, the newer and less well-known disproportionality work is beginning to indicate that the majority of all pollution is clearly not "economically necessary." Instead, the worst problems of pollution and resource exhaustion appear to come from surprisingly tiny fractions of all facilities and companies – many or most of which are in industries where the vast majority of other facilities and companies do not find it "necessary" to produce the same levels of pollution.

(5) *Equity and Inequality as CAUSES of Environmental Problems.* Finally, if most discussions of equity and the environment to date have focused on unequal consequences of environmental harms, another small but growing body of work, this time largely in ecological economics, presents the flip side of the argument. One key component of the common view just noted – namely that prosperity may be one of the best antidotes to environmental problems – is the assumption that people will only care about environmental preservation after they or their grandchildren have become reasonably prosperous. In recent years, however, research has shown that the people in the world's poorest and richest countries tend to report remarkably similar levels of environmental concerns (see e.g. Dunlap, Gallup, & Gallup, 1993). In addition, a new body of work, best represented in this volume in the chapter by Boyce (2008), has begun to pay closer attention to the fact that environmental harm is often inflicted upon the poor by the rich, but rarely inflicted on the rich by the poor. Given that citizens in more egalitarian societies may have greater ability to prevent such harms from being imposed upon them, this work argues that the level of inequality in society may well be one of the best predictors available of how much environmental harm will be created.

A LOOK TO THE FUTURE

To date, the limited level of attention devoted to equity issues has perhaps been clearest in work done outside of the social sciences, but in our view, even social science work on environmental issues has devoted too little attention to the importance of equity and inequality. In an chapter that proved particularly influential in spelling out approaches for analyzing relationships between society and environment, for example, Dunlap and Catton (1983); see also Dunlap (1993) noted the importance of recognizing three "analytically distinguishable functions" of the biophysical environment, reflecting the fact that humans tend to use the environment as (1) a dwelling place, (2) a source of supplies, and (3) a repository for wastes. The typology has been influential in part because it is so simple and so useful. In addition, as noted in later work by Dunlap (1993) and Dunlap and Catton (2002), for example, one factor behind rising awareness of environmental problems, particularly in the context of the social problems literature, is that these three functions tend to be incompatible with one another. In present-day homes, for example, we tend to separate bathrooms from bedrooms from eating places, and more broadly, across many of the world's cultures, the pressure to devote more and more of the finite space available to one such function – for example, waste disposal – is in fact increasingly running into competing demands to use the same spaces for resource supplies or living spaces.

The view that inspired us to work on this volume, however, is that the time has come for the simple Dunlap–Catton typology to be expanded. The three functions of living space, resource supply, and waste sink, need to be joined by a fourth – the use of the biophysical environment as location of and a set of channels for distributing societal inequalities. As the chapters in this volume point out, in other words, the problem is not just that an undifferentiated need for waste disposal sites is increasingly bumping up against an undifferentiated need for living spaces. It is that some social actors are producing more than a proportionate share of harm, and enjoying disproportionate share of benefits – all while asking the policy system to dispose of the wastes and other negative byproducts in or near the "living spaces" of *other* social groups. Ironically, while the "Not In My Back Yard" or "NIMBY" label is often directed at the innocent and predominantly poor bystanders whose neighborhoods have been chosen by corporations and policy actors as the locations for waste-disposal facilities, the people who truly originate and personify the "Not In My Back Yard" pattern are the far more affluent people who own the companies and live in the neighborhoods that appear to benefit from producing those same wastes.

The pattern is all the more remarkable in that it differs so starkly from the insights that are most commonly expressed in those courses on environmental ethics to which we referred in the opening paragraphs of this chapter.

Perhaps the most common of all ethical views – whether we are referring to the views of traditional or indigenous cultures, or to the views of the most sophisticated of modern, Western-trained philosophers (see e.g. Shrader-Frechette, 1981) – is that, if the earth and its resources can legitimately be seen as "belonging" to humans, at all, the resources are ones that are inherently shared. What we generally see in the worlds of economics and policy, however, are nearly the opposite outcomes.

As noted by Freudenburg (2005), for example, what we call oil "production" is actually nothing of the sort. The oil was actually "produced" hundreds of millions of years ago, when no creatures even resembling human beings were present on the earth. What at least some humans do is to remove the ancient and non-renewable resource from the ground and then burn it up. Calling that process "production" is at least remarkable – worthy of being remarked upon.

During all of the time when that "natural" resource is being turned into one of the most valuable commodities on the face of the earth, moreover, humans tend to account for it with high levels of economic and engineering precision. In that accounting, to note an obvious point that is nevertheless worth emphasizing – particularly given the degree to which it differs from what environmental philosophers have urged – the usual tendency is to think of that "natural" resource not as belonging to "nature," or even as something that belongs to all humanity, but to the contrary, as being "private property," over which individuals have "rights." Ironically, it is only when certain humans turn that valuable resource into something we actually do produce – the emissions that come out of an oil refinery or an automobile tailpipe, for example – that our terminology begins to change. These waste products – which, unlike the petroleum, truly are "produced" by humans – suddenly do start to be described by the economic and policy worlds as being "the common legacy of mankind." The magical transformation of language takes place at the precise moment when the emissions are actually "produced" by human activity but no longer desired by the specific humans who enjoyed a profit from producing them (see Freudenburg, 2005 for further discussion).

At least in scholarly and scientific work to date, as this example suggests, the biophysical environment may rarely have been analyzed in terms of inequalities, but it has long been the context and the channel for their distribution. Human uses of the environment even provide the basis for the common expression regarding inequality – living on "the wrong side of the tracks" – which once reflected the fact that coal-burning railroad engines spewed great quantities of cinders and soot, and that those who lived downwind of the tracks tended to receive more than an equal share of the fallout.

As human appetites grow, however, and as the finite limits of the earth become increasingly evident, we fully expect that the connections between equity and the environment will come to take on much greater importance over the years ahead. Partly for that reason, and partly because we know that our own students are often quite curious about what famous authors were thinking when they were producing their important works, we have asked the authors in this volume to do more than merely to summarize the key points of the relevant work and to provide the kinds of citations that will allow interested readers to find more of the work they are summarizing.

As will become clear from reading the chapters themselves, we have instead also explicitly asked the authors to write with "more personality" than is usually expected or even allowed in peer-reviewed journals such as this one. Although this additional request needed to be discussed with the reviewers as well, we specifically asked the authors in this volume to tell readers (including future students) a bit more about how they came to focus on equity questions, how their colleagues have reacted, and more broadly, how the academic and policy worlds have responded to their research, over time, and why. Our reason for doing so relates to our ultimate goal for this volume. We want the volume not just to be a collection of chapters – not even "just" an unusually important collection or chapters – but also to be one of the few publications anywhere that will actually give current and future students some sense of what it was "like" to be a pioneer in doing this area of work.

A Look to this Volume

Inequality in the Experience of Environmental Harm

The chapters themselves come in four sets of two. The first pair of chapters both focus on what is perhaps the best-developed body of research, to date, on issues of equity and the environment, involving *Environmental Justice*. Both of these authors have been central to the growing awareness of (and sophistication in) the environmental justice literature, and both are co-authors of the *Toxic Wastes and Race at Twenty* volume noted earlier. In the first of the two, Paul Mohai (2008) discusses *Equity and the Environmental Justice Debate*. As he notes, work in this area has been the subject of considerable "debate" over the past 20 years or more, but thanks in significant part to his own, carefully performed, quantitative analyses, there is today very little remaining debate among serious researchers about the overall pattern: At least in the United States, there is a strong correlation between the color of a citizen's skin and the likelihood of living in close proximity to toxic and hazardous facilities. Dr. Mohai's chapter offers a particularly helpful introduction to this entire field of work, providing the historical background, noting the emergence of controversy during the 1990s, and then methodically and carefully examining the possible explanations that have been offered to date – ruling out essentially all of the possible excuses that do not reduce down to differential risks for members minority groups in the U.S.

In the next chapter, Robert Bullard (2008) – arguably the one person in all of the social sciences most responsible for convincing the rest of us that this is a topic that deserves greater attention – discusses *Equity, Unnatural Man-Made Disasters, and Race: Why Environmental Justice Matters.* In an especially compelling account, this chapter explores the interweaving of biography and science, noting some of the powerful ways in which Dr. Bullard's own experiences may have helped contribute to the skillful and compelling ways in which he – in conjunction with other respected researchers, some of whom are also authors in this volume – has helped to identify, to shape, and ultimately to define in the eyes of his fellow social scientists, policy-makers, and the public, the entire field of environmental justice research. Dr. Bullard also introduces a theme that is echoed in other chapters, noting that even if persons from a broad range of racial and sociodemographic backgrounds may use a closely comparable set of research tools for *answering* research questions, personal backgrounds can be a powerful factor in *selecting* the research questions we attempt to answer.

Given the power of personal background in shaping the questions we choose to study – and given the number of decades over which the vast majority of all active researchers in the U.S. have been both white and male – it is important to realize that the contours of the scientifically known world might look very different if the composition of the scientific community had not been so low in diversity, for so long. The importance of drawing on a more diverse set of research questions and insights, however, is one that has at least equally powerful implications for the future – a point that is examined in the next pair of chapters.

Race and Gender in the Management and Analysis of Environmental Problems
Continuing the theme of forthrightly examining the interaction of personal biography with the development of research, with a special emphasis on the

selection of research topics, Dorceta Taylor (2008) discusses *Diversity and the Environment: Myth-Making and the Status of Minorities in the Field.* As Dr. Taylor spells out, her own experiences and background were particularly influential in leading her to question, empirically, the usual arguments for why so few women and members of minority groups, even today, have been hired for work in environmental fields. Scientifically speaking, this turns out to be an example of why greater diversity among scientists is important. As she notes, the "usual arguments" may rarely have been questioned by white male scientists in the past, but when examined directly, they ultimately prove to be myths.

Summarizing the most extensive body of evidence available on the topic, Dr. Taylor also explores one of the most obvious ways in which "mainstream" environmental organizations could improve their sensitivity to the concerns of women and of the ever-growing segments of the population whose ancestors came from areas other than Europe, namely hiring a more diverse workforce. As she reports, the usual excuses – holding that women and minorities are uninterested in the environment, unwilling to work for environmental organizations, desirous of salaries that are too high, and/or lacking in qualifications – simply do not hold water. As should be clear to anyone who actually teaches in college-level environmental programs today, where ever-growing fractions of the top students are females and/or persons from minority backgrounds, there are growing numbers of highly qualified women and minority students with relevant degrees. In addition, women and minority students are desirous of jobs in environmental organizations, and their salary expectations are within the range of what is currently being paid. All in all, her findings suggest that a more serious problem may be the fact that environmental organizations still tend to rely on network ties and informal mechanisms such as word-of-mouth techniques to recruit their staffs. Although many of the organizations in question appear not to have thought about the problem in any detail, referral networks tend to replicate the demographic characteristics of the incumbent workforce, preventing the organization from learning as much about minority candidates who are at least equally qualified.

In the next chapter, Anne Statham and Christine Evans (2008) further explore the ways in which the development of a more diverse body of scientists might reshape the contours of the insights being produced in the future. Their chapter focuses on *Extending the Reach of Ecofeminism: A Framework of Social Science and Natural Science Careers.* Introducing another theme that will be echoed in subsequent chapters, these authors explore the transcending of another divide – the veritable chasm that has for too long tended to separate those who study humans from those who study hardware and habitat, or in more prosaic terms, the boundary between the scholars and scientists who study humans and our cultures, on the one hand, versus those who study the physical and biological characteristics of our surroundings, on the other.

As Drs. Statham and Evans note, the field known as "Ecofeminism" is a relatively new but also a large and complex one. As their chapter spells out in a clearly personal fashion, the ways in which this new form of thinking may influence a scientist's orientations and insights can be shaped in powerful ways not just by sociodemographic characteristics, but also by the field of science in which one works. Just as having a more diverse representation of humanity involved in scientific and scholarly pursuits may enrich those pursuits, in other words, the thoughtful consideration of a more diverse representation of *academic* perspectives may add a still-more-diverse form of enrichment – even to people who may look much more similar in terms of skin color, age, and sex.

Equity and Inequality in Extreme Environments
The next two chapters explore further the challenges and rewards of attempting to build the kinds of interdisciplinary collaborations introduced by Drs. Statham and Evans, seeking to address in an increasingly integrated way the kinds of topics that until relatively recently were much more likely to be addressed separately by social versus biophysical scientists. It may be only a coincidence, but both of the next two chapters also explore the special concerns that emerge when issues of Equity and Inequality are considered in "Extreme Environments" – some of the hottest and coldest places on the planet.

As suggested by the number of people involved in the first of these next two chapters, the authors involved – Sharon L. Harlan, Anthony J. Brazel, G. Darrel Jenerette, Nancy S. Jones, Larissa Larsen, Lela Prashad, and William L. Stefanov – have worked as a deliberately interdisciplinary team. As such, this team has experienced the challenges as well as the opportunities of learning from interdisciplinary colleagues, and the impressive results of their complex interactions are reported in their chapter, *In the Shade of Affluence: The Inequitable Distribution of the Urban Heat Island*. Intriguingly, the chapter helps to illustrate one of the very arguments it makes, offering the kinds of insights that are rarely seen in "disciplinary" journals – whether those journals focus on biophysical or on social sciences – but that are instead sometimes made possible by the fact that the authors are working across disciplinary boundaries. As the chapter notes, our understanding of the world would be improved if the academic world offered more inducements, and fewer impediments, to the carrying out of such interdisciplinary collaboration.

If Dr. Harlan and her colleagues are examining one of the hottest and most urban of populations in North America, the next chapter, by Bruce Forbes (2008), focuses on one of the coldest and least urban of regions on the planet. Dr. Forbes addresses *Equity, Vulnerability and Resilience in Social-Ecological Systems*, offering "A Contemporary Example from the Russian Arctic." If the chapter by Harlan and her colleagues is an illustration of what can be done by an interdisciplinary team, similarly, the chapter by Dr. Forbes provides an illustration of what can be done by an interdisciplinary *individual* – one whose subject matter inherently crosses national as well as disciplinary boundaries.

Dr. Forbes focuses on issues of equity in the face of climatic as well as land-use changes among reindeer-herding peoples, particularly in Russia's northwestern Siberia. As he points out, issues of "sustainable development" take on special challenges in northern regions, particularly given that the disruptive effects of global warming are being felt with special force in the far North. Still, given that the human realities of the Russian far north are starkly different even from those of the northern people of Europe and North America who have been studied in greater detail, Forbes argues persuasively that it would be a mistake simply to assume that findings from other northern peoples will apply well to the reindeer herders of the Russian far north.

Inequality and Equity as Predictors of Environmental Harm
The last two chapters in this volume offer not so much the summing up of well-known areas of research, but the opening up of relatively new ones. In a sense, both of these last two chapters also reverse the direction in which cause and effect are considered, dealing with inequality not so much as the consequence of environmental disruption but as the cause of it.

The chapter by Lisa Berry (2008) assesses *Disproportionality Inequality in the Creation of Environmental Damage*. As she notes, although Garrett Hardin's metaphor of "the Tragedy of the Commons" (Hardin, 1968) has proved highly influential in thinking about the interactions between humans and habitat, that influence has not been universally beneficial. Instead, as she spells out, growing bodies of evidence suggest that, rather than being the result of all of us acting in a common or shared way, many of the most severe environmental problems result from grossly disproportionate impacts from a surprisingly small fraction of all economic activity. She goes on to note that, rather than representing the most vital and dynamic of new or growing economic sectors, the disproportionately polluting activities tend to be associated with older, relatively low-technology industries.

As Berry points out, moreover, the accumulating findings suggest that it would be misleading even to think of these disproportionately polluting activities as being "representative" of their own industries. Instead she explains, a substantial fraction of all toxic emissions come from facilities that are responsible for far more pollution than are other facilities in the same industries. She notes, for example, that for the most toxic single industry in the United States in terms of estimated risks – the primary nonferrous metal industry – if the "top" or most heavily polluting 10% of the firms in the industry were to reduce their emissions *per dollar* to the median for the same industry, the net result would be dramatic: The emissions from this most toxic of American industries would go down by 83.5%. More broadly, while past research has stubbornly focused on "average" polluters or "overall" impacts, Berry notes that such a focus routinely overestimates the actual impacts associated with the vast majority of firms in an industry – or an economy – while simultaneously failing to focus on those few firms where reductions could make the greatest difference.

The closing chapter, by James K. Boyce (2008) asks the simple but provocative question, *Is Inequality Bad for the Environment?* In particular, he focuses on a pair of hypotheses. The first one challenges the expectation that environmental harm will be good for the well-being of ordinary citizens. His chapter argues, instead, that those who are relatively wealthy and powerful will tend to benefit disproportionately from any economic activities, specifically including the activities that generate environmental harm, while the relatively poor and powerless will tend to bear a disproportionate share of the environmental costs. The second hypothesis takes the argument a step further, noting that the level of inequality in a given society is likely to shape not just the distribution, but the total magnitude of environmental harm that will be created. The first of these two hypotheses may seem intuitively plausible – although his chapter illustrates that it may be correct for a larger number of reasons than might first come to mind. His second hypothesis, by contrast, might seem less intuitively plausible – in fact, it was considered by many to be a radical claim when he first proposed it – but as he spells out in his chapter, findings to date have tended to support both hypotheses. Societies with wider inequalities of wealth and power, for example, do appear to suffer higher levels of environmental harm – *overall*, and not just for their poorest or least powerful members – while societies having lower levels of economic and political inequality will also tend to have lower levels of environmental harm.

One study he cites in his chapter, for example, found that higher levels of racial and ethnic segregation were associated with worse environmental and health outcomes for all groups, and not just for members of minority groups.

In a fitting close to the volume, he shows that the answer to the question asked by his chapter is a positive one. Inequality is bad not just for victims, but for the environment. The way in which we treat the environment depends in part on how we treat each other – and it has further consequences for the way we treat one another, in turn. It is a conclusion that is relevant both to the levels of equity and inequality at any given time, and to inequality across time. As Dr. Boyce points out, humans may be unique in the levels of environmental disruption that we have created, but we are also unique in our ability to anticipate future problems and to act to avoid them.

CONCLUSION

The chapter in this volume by Statham and Evans reports that Dr. Evans still remembers thinking – after she was turned down for an internship with a federal resource agency because she was a married woman – that "Feminism is about justice, and science is about facts." If anything, the chapters in this volume, collectively, tell a different story. As Dr. Evans later concluded herself, such a neat but artificial separation may be one of the problems of the ways in which we have thought about connections between people and nature in the past. Levels of justice and inequity are also facts – facts that need to be considered more squarely by all scientists in the future, including those who work in the biophysical as well as the social sciences.

As the concluding chapter by Boyce points out, the accumulated experience of humankind certainly provides no guarantees that issues of equity and inequality will be handled in a more thoughtful manner in the future – let alone one that would be more egalitarian – and yet there may well be reasons for optimism. Just 250 years ago, most of the people of the world were "subjects" who lived under monarchs, and just 150 years ago, it was still legal for one group of human beings to "own" another group of human beings – slaves – in the United States. In this same country, it was less than 100 years ago when women finally won the constitutional right to vote, and as recently as 50 years ago, "separate but (supposedly) equal" racially segregated schools were still supported by the law of the land. Thus, while there is very little likelihood that inequalities will disappear from the earth in the near future – either in terms of human uses of the environment or in any other respect – the longer-term trends do appear to provide reasons for hope.

Those changes, however, were not automatic; they were the result, in part, of widespread and yet focused attention. The key first step, accordingly, is that issues of equity and inequality need to become the subjects of focused and careful attention, not ongoing neglect. One of the preconditions for more enlightened approaches, in short, may be the very process of bringing uncomfortable facts to light in the first place.

Equity issues were generally overlooked by most of the "mainstream" work on environmental topics during the closing decades of the 20th century, but the more attention they have been getting, the more important we are coming to see they are. Growing evidence suggests that harming the environment potentially harms other humans, both now and in the future, and that harming other humans may harm the environment, with consequences that can come back in turn to haunt not merely those who were the immediate victims, but all of us.

For the 21st century, then, issues of equity and inequality need to become the focus of far more attention, both in studies that deal with the earth's environment and those that do not. The research summarized in this volume, for all of its insights to date, needs to be recognized as being far from a completed body of work; instead, it is the highly promising start of a much larger body of insights and accomplishments that still need to come in the future. There is no better time to start than now. Rather than remaining relegated to the academic equivalent of some small pigeonhole, issues of equity and inequality need to become the focus of more than just this volume. They need to become central to the study of the connections between humans and the habitat that we share with all other life on earth.

REFERENCES

Berry, L. (2008). Disproportionality and inequality in the creation of environmental damage. *Research in Social Problems and Public Policy, 15* (In press).

Boyce, J. K. (2008). Is inequality bad for the environment? *Research in Social Problems and Public Policy, 15* (In press).

Bullard, R. (2008). Equity, unnatural man-made disasters, and race: Why environmental justice matters. *Research in Social Problems and Public Policy, 15* (In press).

Bullard, R. D., Mohai, P., Saha, R., & Wright, B. (2007). *Toxic wastes and race at twenty 1987–2007: Grassroots struggles to dismantle environmental racism in the United States.* Cleveland, OH: United Church of Christ Justice and Witness Ministry.

Commission for Racial Justice, United Church of Christ. (1987). *Toxic wastes and race in the United States.* New York: United Church of Christ.

Dunlap, R. E. (1993). From environmental to ecological problems. In: C. Calhoun & G. Ritzer (Eds), *Social problems* (pp. 707–738). New York: McGraw Hill.

Dunlap, R. E., & Catton, W. R., Jr. (1983). What environmental sociologists have in common (Whether concerned with "built" or "natural" environments). *Sociological Inquiry, 53*(Spring), 113–135.

Dunlap, R. E., & Catton, W. R., Jr. (2002). Which function (s) of the environment do we study? A comparison of environmental and natural resource sociology. *Society and Natural Resources, 15*, 239–249.

Dunlap, R. E., Gallup, G. H., Jr., & Gallup, A. M. (1993). *Health of the planet: Results of a 1992 International Environmental Opinion Survey of Citizens in 24 nations.* Princeton, NJ: The George H. Gallup International Institute.

Forbes, B. (2008). Equity, vulnerability and resilience in social-ecological systems: A contemporary example from the Russian Arctic. *Research in Social Problems and Public Policy, 15* (In press).

Freudenburg, W. R. (2005). Privileged access, privileged accounts: Toward a socially structured theory of resources and discourses. *Social Forces, 94*(1), 89–114.

Hardin, G. (1968). The tragedy of the commons. *Science, 162* (December), 1243–1248.

Harlan, S. L., Brazel, A. J., Jenerette, G. D., Jones, N. S., Larsen, L., Prashad, L., & Stefanov, W. L. (2008). In the shade of affluence: The inequitable distribution of the urban heat island. *Research in Social Problems and Public Policy, 15* (In press).

Mohai, P. (2008). Equity and the environmental justice debate. *Research in Social Problems and Public Policy, 15* (In press).

Niazi, T. (1999). The questionable success of economic growth in Pakistan. *Research in Social Problems and Public Policy, 7*, 199–216.

Orr, D. W. (1992). *Ecological literacy: Education and the transition to a postmodern world.* Albany: State University of New York Press.

Romero, A., & Silveri, P. (2006). Not all are created equal: An analysis of the environmental programs/departments in U.S. academic institutions from 1900 until May 2005. *Journal of Integrative Biology, 1*(1), 1–15. Also available on-line: http://clt.astate.edu/electronicjournal/Articles.htm

Shrader-Frechette, K. (1981). *Environmental ethics.* Pacific Grove, CA: Boxwood Press.

Statham, A., & Evans, C. (2008). Extending the reach of ecofeminism: A framework of social science and natural science careers. *Research in Social Problems and Public Policy, 15* (In press).

Taylor, D. (2008). Diversity and the environment: Myth-making and the status of minorities in the field. *Research in Social Problems and Public Policy, 15* (In press).

World Commission on Environment and Development (Gro Harlem Brundtland, Chair). (1987). *Our common future.* Oxford: Oxford University Press.

PART I:
ENVIRONMENTAL JUSTICE

EQUITY AND THE ENVIRONMENTAL JUSTICE DEBATE

Paul Mohai

ABSTRACT

This article takes an autobiographical approach in describing the evolution of the equity and environmental justice debate. The intent is not only to provide a historical approach in identifying the emerging research and policy questions, but also to describe the author's own scholarly growth in studying them.

The environmental justice movement has spurred much academic interest and policy debates about the existence, causes, and solutions to environmental inequalities based on racial and socioeconomic factors. The earliest research attempted to determine the existence and magnitude of such disparities. Evidence of the existence of such disparities has been enough to spur government action. Although some researchers have questioned the existence and seriousness of such disparities, systematic reviews have shown that the weight of the evidence tends to support the claims of the movement. Nevertheless, challenges to the claims of environmental justice activists, supported at times by contrary research evidence, have stimulated a great deal of attention to questions about the validity of the methodologies for

Equity and the Environment
Research in Social Problems and Public Policy, Volume 15, 21–49
Copyright © 2008 by Elsevier Ltd.
ISSN: 0196-1152/doi:10.1016/S0196-1152(07)15001-8

assessing racial and socioeconomic disparities around environmentally hazardous sites and whether the evidence produced by them conclusively demonstrates that disparities exist. There has also been much interest in understanding how racial and socioeconomic disparities in the distribution of environmental hazards come about, and in probing the economic, health, and other quality of life impacts associated with living near environmentally burdened sites. The author has been and is currently involved in all aspects of the research seeking answers to these questions. This article describes the author's involvement and in so doing attempts to identify and discuss the important research issues focused on environmental justice and their policy implications.

RACE AND CONCERN FOR THE ENVIRONMENT: DISPELLING OLD MYTHS

Initially I became involved as a researcher in the area of what is now termed "environmental justice" because of my interest in better understanding the environmental concerns and attitudes of African Americans and other people of color. This interest emerged in the late 1980s. Earlier research found that there was little empirical evidence to show that working class people or poor people are less concerned about the environment than others (Buttel & Flinn, 1978; Mitchell, 1979; Mohai, 1985; Van Liere & Dunlap, 1980). I began wondering whether the conventional wisdom of the day, that African Americans are unconcerned about the environment, at least that they are less concerned than white Americans, was also unfounded. Questions about this conventional wisdom were already being raised in the 1980s by African American scholars, such as Professors Robert D. Bullard, Beverly Wright, and Dorceta Taylor (Bullard & Wright, 1986, 1987a, 1987b; Taylor, 1989). I had at my disposal data from a very large national survey of environmental attitudes, based on over 7000 face-to-face interviews conducted by Louis Harris and associates. Almost 600 African Americans were interviewed in this survey, making it the largest national survey of African American environmental attitudes ever conducted.

I began working in earnest on researching African American environmental attitudes in 1987 and was interested in developing some hypotheses for my analysis (see Mohai, 1990 for the results; see also Mohai & Bryant, 1998; Mohai & Kershner, 2002; Mohai, 2003 for updates to that early work).[1] That was the year I began as an Assistant Professor at the University of Michigan and where I met my colleague Professor Bunyan

Bryant, the first African American environmental studies professor I ever met. I told him about my research interests and he pointed me to the recently released report, *Toxic Wastes and Race in the United States*, published by the Commission for Racial Justice of the United Church of Christ (UCC, 1987). This report had a major influence on me. Until that time, I had not stopped to consider that environmental problems may be disproportionately burdening certain groups in society over others. Public discourse at that time generally held that environmental problems are everyone's problems and that everyone has an equal stake in solving them. However, the UCC report was suggesting that this was not the case – that instead, racial and ethnic minorities and the poor may be at significantly greater risk, and thus that they may have a greater stake than others in solving environmental problems. The most compelling aspects of the report were the findings that, nationally, the minority percentages of zip code areas with hazardous waste facilities were twice as great as those of zip code areas without facilities – with the percentages being three times greater for zip code areas with two or more facilities. Furthermore, multivariate statistical analyses revealed that, among a variety of variables related to facility location, including mean property values, mean incomes, and the presence of abandoned and uncontrolled hazardous waste, the minority percentage of the zip code areas was found to be the best predictor of where hazardous waste facilities are located. I was also struck by the fact that the reason that this report existed was in response to a growing movement by people of color to protest the environmental burdens in their communities and to seek remedies to their perceptions of unfair and unequal treatment by industry and government.

I joined forces in 1987 with Professor Bryant in an effort to understand these issues better. We began with three objectives. One was to determine whether other quantitative studies assessing environmental disparities existed, and whether these studies produced similar findings. This effort resulted, to our knowledge, in the first systematic review of such quantitative studies (see Mohai & Bryant, 1992). A second objective was to conduct a comprehensive study of the distribution of environmental hazards in the Detroit metropolitan area and to examine how such a distribution affected African American environmental attitudes (see Mohai & Bryant, 1992, 1998). Our third objective was to organize a conference bringing together all the researchers around the nation that we could identify examining racial and socioeconomic disparities in the distribution of environmental hazards, asking them to present and discuss the findings and implications of their most recent research. In doing so, we sought to increase the visibility of this

issue among other academic researchers as well as to get the attention of
state and national policy makers, especially that of the U.S. Environmental
Protection Agency (EPA). We named our conference the "Michigan
Conference on Race and the Incidence of Environmental Hazards" (for
discussions of the role of the Michigan Conference, see the chapters by
Bullard and Taylor in this volume; see also Bryant & Mohai, 1992a, 1992b).

THE 1990 MICHIGAN CONFERENCE

One of the outcomes of the Conference was the decision by the participants
to draft a letter to then EPA Administrator William Reilly requesting a
meeting with him and his staff to talk about the evidence pertaining to
environmental inequalities, and to explore what the agency might do to
address this problem. Our efforts were largely successful. Representatives
from the Conference were invited to meet with Administrator Reilly that
September (we were subsequently dubbed by EPA "The Michigan
Coalition" and included Bunyan Bryant, Robert Bullard, Benjamin Chavis,
Michel Gelobter, David Hahn-Baker, Charles Lee, Paul Mohai, and Beverly
Wright). One of the most important outcomes of this meeting was the
decision by Administrator Reilly to create an internal EPA working group
(the "Environmental Equity Workgroup") to investigate the evidence and
draft a set of proposals to address environmental inequalities. (Later, EPA
created an Office of Environmental Equity, which was renamed the "Office
of Environmental Justice" under the Clinton Administration). In 1992,
EPA's Environmental Equity Workgroup produced a report entitled
Environmental Equity: Reducing Risk for All Communities. Although, the
report's draft proposals were heavily criticized for not going far enough in
addressing the problem of environmental injustice (see Volume II of that
report), the report was nevertheless significant: as the first official acknowl-
edgement of the problem by the agency and federal government, the report
raised the visibility of this issue and legitimized environmental injustice as an
issue warranting further federal attention. Indeed, subsequent to the report,
a series of hearings on environmental justice were held in the U.S. Congress,
and several bills were introduced. Although none passed, the increasing
attention and efforts by activists, academics, and the Congressional Black
Caucus and other supporters eventually led President Clinton in 1994 to
issue Environmental Justice Executive Order 12898, calling upon on all the
agencies in the federal government, not just the EPA, to make environmental justice a priority in their rule making.

EXAMINING THE EVIDENCE

At the same time Professor Bryant and I were planning the 1990 Michigan Conference, we wanted to see if there was corroborating evidence to that presented in the UCC report. As mentioned above, one of our efforts was to search the literature for other quantitative studies focused on environmental disparities. This was a difficult task, as not much had been published on the topic at that time, and electronic databases that currently facilitate the searching of the academic literature were not widely available then. I started with the reference list in the UCC report. I also enlisted the aid of search specialists at the University of Michigan Library. I was struck that the use of key words such as "race" and "environment" brought up mostly the articles that Professor Robert D. Bullard and his colleague and co-author Professor Beverly Wright had written. I found that Professor Bullard had not only done one of the earliest studies of environmental disparities (Bullard, 1983), but also that he and his co-author Professor Wright had already written extensively in the 1980s about the emerging issues of environmental injustices and their likely causes (see, e.g., Bullard & Wright, 1986, 1987b). Yet broader academic attention had not yet emerged on this topic. Eventually, I contacted Professor Bullard to find out more about what had been researched and written about this topic.

Through his help, the University of Michigan Library, and other sources, I found 15 empirical studies that had examined racial and socioeconomic disparities in the distribution of environmental hazards of a wide variety (see Table 1). As mentioned above, I was specifically interested in determining whether the findings of other empirical studies pointed in the same direction as that of the UCC report. Specifically, I wanted to know (1) whether other quantitative studies found significant racial and socio-economic disparities in the distribution of environmental hazards and (2) whether racial disparities tended to be more important than income and other socioeconomic disparities. All 15 studies I reviewed found either significant racial or socioeconomic disparities, and where it was possible to weigh the relative importance of race or income in predicting where environmental hazards were located, in five out of eight studies race was found to be the most important predictor (see Table 1 and Mohai & Bryant, 1992). Given that these studies varied in their geographic scope (some studied specific metropolitan areas, others were regional, while still others were national in scope), examined a wide variety of hazards (including air pollution, garbage dumps, and hazardous waste sites), and varied in their methodological approaches, the robustness of the pattern appeared to offer

Table 1. Empirical Studies, 1992 and Earlier, Providing Systematic
Evidence Regarding the Burden of Environmental Hazards by Income
and Race.

Study	Hazard	Focus of Study	Distribution Inequitable by Income?	Distribution Inequitable by Race?	Income or Race More Important?
CEQ (1971)	Air pollution	Urban area	Yes	NA	NA
Freeman (1972)	Air pollution	Urban areas	Yes	Yes	Race
Zupan (1973)	Air pollution	Urban area	Yes	NA	NA
Harrison (1975)	Air pollution	Urban areas	Yes	NA	NA
		Nation	No	NA	NA
Kruvant (1975)	Air pollution	Urban area	Yes	Yes	Income
Burch (1976)	Air pollution	Urban area	Yes	No	Income
Berry et al. (1977)	Air pollution	Urban areas	Yes	Yes	NA
	Solid waste	Urban areas	Yes	Yes	NA
	Noise	Urban areas	Yes	Yes	NA
	Pesticide poisoning	Urban areas	Yes	Yes	NA
	Rat bite risk	Urban areas	Yes	Yes	NA
Handy (1977)	Air pollution	Urban area	Yes	NA	NA
Asch and Seneca (1978)	Air pollution	Urban areas	Yes	Yes	Income
Gianessi, Peskin, and Wolff (1979)	Air pollution	Nation	No	Yes	Race
Bullard (1983)	Solid waste	Urban area	NA	Yes	NA
U.S. GAO (1983)	Hazardous waste	Southern region	Yes	Yes	NA
United Church of Christ (1987)	Hazardous waste	Nation	Yes	Yes	Race
Gelobter (1988, 1992)	Air pollution	Urban areas	Yes	Yes	Race
		Nation	No	Yes	Race
West et al. (1992)	Toxic fish consumption	State	No	Yes	Race

Note: NA = Not applicable.
Source: Adapted from Mohai and Bryant (1992).

solid confirmation of the UCC report's findings. Since our review, two additional systematic reviews of the empirical evidence have been conducted, both of which have arrived at similar findings and conclusions (see Goldman, 1994; Ringquist, 2005).

In addition to conducting a review of existing evidence, Professor Bryant and I wanted to conduct a new study, so we began working on a proposal to examine environmental disparities in the Detroit metropolitan area. We also wanted to know how such disparities affect racial differences in environmental awareness and concerns. Although prior environmental inequality studies employed census data at that time, it was apparent that survey data would be especially ideal for achieving our two objectives. We also realized that the University of Michigan's annual Detroit Area Study (DAS) would be an ideal vehicle for conducting such a study, so we submitted our proposal for consideration for the 1990 DAS and were successful. We not only developed a comprehensive survey to compare African American and white American environmental attitudes, but we mapped the locations of the homes of all the 793 1990 DAS respondents as well as all the locations of the hundreds of polluting industrial facilities, commercial hazardous waste facilities, and abandoned hazardous waste sites in the Detroit area. We measured the distances between respondents and the hazardous sites. Because of the UCC report's focus on commercial hazardous waste facilities, we were specifically interested in examining the proportion of people of color living near such facilities in the Detroit area. We found that, with increasing proximity to the hazardous waste facilities, the proportions of African American residents relative to white residents increased (Mohai & Bryant, 1992). Through multivariate statistical analyses, we also found that the race of the respondents was more important than their incomes in predicting their proximity to these facilities. Subsequently, with the help of graduate students in the School of Natural Resources and Environment, we completed our analyses of racial disparities in the distribution of a wide range of environmentally hazardous sites (e.g., polluting industrial facilities and abandoned hazardous waste sites), and found the patterns consistent with what we initially found for commercial hazardous waste facilities (see Table 2 and Mohai & Bryant, 1998). That is, larger proportions of the African American population in the Detroit metropolitan area relative to the white population were found living near hazardous sites and areas of poor environmental quality. The results of our studies thus appeared to corroborate the findings of the UCC report and earlier environmental inequality studies.

NEW RESEARCH

The UCC's (1987) *Toxic Wastes and Race in the United States*, the 1990 Michigan Conference, Bullard's (1990) *Dumping in Dixie: Race, Class, and Environmental Quality*, Bryant and Mohai's (1992a) *Race and the Incidence of Environmental Hazards: A Time for Discourse* (containing the published

Table 2.   Percent of African Americans and Percent of Whites in the
Detroit Metropolitan Area Living in Proximity to a Potential
Environmental Hazard or Poor Environmental Quality.

	African Americans	Whites	Difference
Proximity to polluting industrial facility	$n = 130$	$n = 629$	
(1) Further than 1.5 miles	26.7	50.1	−23.4
(2) Between 1 and 1.5 miles	22.3	18.4	3.9
(3) Within 1 mile	51	31.5	19.5
$\chi^2 = 25.63^{***}$			
Proximity to commercial hazardous waste facility	$n = 130$	$n = 629$	
(1) Further than 1.5 miles	64.5	90.3	−25.8
(2) Between 1 and 1.5 miles	19.6	6.7	12.9
(3) Within 1 mile	15.9	3	12.9
$\chi^2 = 63.18^{***}$			
Proximity to uncontrolled hazardous waste site	$n = 130$	$n = 629$	
(1) Further than 1.5 miles	26.1	46.3	−20.3
(2) Between 1 and 1.5 miles	26.7	14	12.7
(3) Within 1 mile	47.2	39.7	7.5
$\chi^2 = 22.45^{***}$			
Commercial or industrial structure on resident's block	$n = 130$	$n = 624$	
(0) No	91.7	93	−1.3
(1) Yes	8.3	7	1.3
$\chi^2 = 0.28$			
Vacant buildings on resident's block	$n = 130$	$n = 624$	
(0) No	75.7	97	−21.3
(1) Yes	24.3	3	21.3
$\chi^2 = 79.25^{***}$			
Upkeep of structures on resident's block	$n = 130$	$n = 619$	
(1) Very well	22.6	66.2	−43.6
(2) Mixed	66.6	32.4	34.2
(3) Poorly	7.4	1	6.4
(4) Very poorly	3.4	0.3	3.1
$\chi^2 = 100.03^{***}$			
Upkeep of sidewalks and yards on resident's block	$n = 130$	$n = 621$	
(1) Very well	24.3	55.3	−31
(2) Mixed	60.5	41.4	19.1
(3) Poorly	11	3.2	7.8
(4) Very poorly	4.1	0.0	4.1
$\chi^2 = 67.54^{***}$			

Note: n = Weighted sample size.
Source: Adapted from Mohai and Bryant (1998).
$^{***}p < 0.001$.

versions of the papers presented at the 1990 Michigan Conference), EPA's (1992) *Environmental Equity: Reducing Risk for All Communities*, as well as many other articles, hearings, and conferences on environmental justice led in the 1990s to a flurry of new academic research in a wide variety of disciplines. Most of this new research tended to support the findings of the UCC report and earlier environmental inequality studies and the claims of environmental justice activists. However, a serious challenge to the claims that environmental hazards are distributed disproportionately by race and socioeconomic status emerged in 1994 with the publication of an article in the prestigious journal, *Demography*, by researchers at the Social and Demographic Research Institute (SADRI) at the University of Massachusetts (Anderton, Anderson, Oakes, & Fraser, 1994).

The researchers claimed in this chapter to have replicated the UCC study using 1980 census data, as the UCC study had done, and found contrary findings, i.e., that no statistically significant racial disparities existed in the distribution of commercial hazardous waste facilities. The reason for the contrary findings, the researchers claimed, was in their use of census tracts rather than zip code areas as their units of analysis. The SADRI researchers argued that because census tracts are generally smaller than zip code areas, they are less likely to lead to "aggregation errors" or "ecological fallacies," defined as "reaching conclusions from a larger unit of analysis that do not hold true in analyses of smaller, more refined units" (Anderton et al., 1994, p. 232). This article and subsequent articles by the SADRI researchers (Anderton, Oakes, & Egan, 1997; Oakes, Anderton, & Anderson, 1996; Davidson & Anderton, 2000) led to the questioning of earlier claims about the existence of environmental disparities, especially by race and ethnicity, and motivated a hard look at the methodologies used in conducting environmental inequality analyses.

QUESTIONS ABOUT METHODOLOGY

Although the SADRI research provided added motivation, I had already begun to think about methodological issues in conducting environmental inequality research even before that time. Because most of the research conducted before the 1990 DAS relied on census data, and Professor Bryant and I wanted to conduct an environmental inequality analysis using survey data, it caused me to think about the different methodological issues involved. Rather than two-dimensional space, such as that represented by zip code areas and census tracts, respondents in a survey are geographic points. It seemed logical that, in determining whether or not people of color and poor people disproportionately reside near environmentally hazardous sites, the distance of the respondents to the sites should be measured, and race and socioeconomic characteristics should be correlated with these distances (Mohai & Bryant, 1992, 1998). Yet in the studies using census data, the distance of zip code areas and census tracts to environmentally hazardous sites had still to be considered.

Rather than accounting for distance between census units and hazardous sites, the approach used in the UCC report and SADRI studies, and indeed in most environmental inequality research, has been what has been termed "spatial coincidence" (McMaster, Leitner, & Sheppard, 1997) or "unit-hazard coincidence" (Mohai & Saha, 2006, 2007). In applying this approach, the researcher selects a pre-defined geographic unit, such as tracts or zip code areas, and determines which of the units contain or "host" the hazard of interest and which units do not. The demographic characteristics of the host and non-host units are then compared. Although there is no effort to determine the precise location of the hazard within the unit, it is assumed that people residing in the host units live closer to the hazard under investigation than people living in the non-host units.

I first began to realize that there might be problems with this assumption after conducting in 1992 an analysis for plaintiffs in a lawsuit concerning the location of two hazardous waste landfills in the Houston metropolitan area. As an expert witness in this case, I was asked to determine whether the landfills were disproportionately placed where minority residents were living. To determine an answer to this question, I employed two methodological innovations that had not previously been used in environmental inequality research using census data. One was to contrast the demographic characteristics of the census tracts containing the hazardous waste landfills with the other census tracts in the Houston metropolitan area just *before* the landfills were sited and then to track demographic changes *after* siting. The second was to analyze the demographic characteristics of the non-host tracts *adjacent* to the host tracts separately from the non-host tracts farther away. I found that the host tracts were disproportionately nonwhite, even before the landfills were sited. I also found in this work, and in subsequent analyses of hazardous waste facilities in the Detroit metropolitan area, that the demographic characteristics of adjacent non-host tracts were likely to be more similar to the host tracts than to non-host tracts farther away. Although these results admittedly came from a limited sample, they caused me to wonder whether the proximity of non-host tracts near hazardous sites mattered in other cases, and whether the assumptions of the unit-hazard coincidence method may be leading to an under-estimation of the degree of racial and socioeconomic disparities around environmentally hazardous sites.

The publication of the SADRI article in 1994 and its contrasting findings and conclusions with those of the UCC study, caused me to look even deeper into the methodological questions involving environmental justice research (Mohai, 1995). Given that both studies claimed similar objectives, were of national scope, focused on the same type of potential environmental hazard (i.e., commercial hazardous waste facilities), and went as far as using the same census decade (1980), it seemed odd to me that they would come to such strikingly contrasting findings and conclusions. The SADRI research-ers claimed that the difference had to do with their use of census tracts, a smaller unit of analysis than zip code areas, and a unit less subject to "aggregation errors" and "ecological fallacies." However, was that the only difference?

I decided to give both studies a thorough review and to note all the differences between them that I could find. It became apparent that the two studies differed in more ways than just in their selection of the units of analysis. The two studies also differed in the way that the comparison groups of non-host units were identified. The UCC study used *all* non-host zip code areas in the U.S., whether in metropolitan or non-metropolitan areas, for the comparison group; in the SADRI study, only non-host census tracts in metropolitan areas already containing a hazardous waste facility were considered for the comparison group. For the SADRI study, in short – unlike the UCC study – non-metropolitan areas were unrepresented, as were all metropolitan areas currently without hazardous waste facilities.

Given that the comparison groups of non-host units were constructed differently for the two studies, I wondered if that might also have had an impact on the differing findings. I decided to compare the people of color percentages in the host units and non-host units of the respective studies. At first this presented a difficulty, because the UCC study reported percentages for all racial and ethnic categories combined, while the SADRI study reported percentages for African Americans and Hispanics separately. Because African Americans and Hispanics made up nearly all of nonwhites in 1980 (97.7%) and because the overlap between the two categories was so small (less than 1.0%), I decided simply to add the African American and Hispanic percentages in the SADRI study and compare the result against the minority percentages presented in the UCC report (Mohai, 1995). The comparisons were very revealing. They appeared to contradict the implications of the SADRI researchers that geographic units larger than

Table 3.   Average People of Color Percentages of Areas Hosting
Commercial Hazardous Waste Facilities and the Respective Control
Populations Used in the UCC and SADRI Studies.

	UCC Study (Zip Code Areas)	SADRI Study (Census Tracts)	SADRI Study (Host Tracts and Tracts within 2.5-mile Radius Aggregated)	SADRI Study – 25 Largest Metropolitan Areas (Host Tracts and Tracts within 2.5-mile Radius Aggregated)
Average people of color percentage in host areas	25	24	35	42
Average people of color percentage in respective control populations	12	23	21	24

Source: Adapted from Mohai (1995).

census tracts (i.e., zip code areas) lead to higher minority estimates around
hazardous waste facilities because of aggregation errors. Rather than finding
that the average minority percentage in zip code areas hosting hazardous
waste facilities to be higher, I found it to be nearly *identical* to the average
minority percentage in tracts hosting such facilities (25% vs. 24%,
respectively; see Table 3). Where the real discrepancy in the results occurred
was in the estimates of the minority percentages of the *comparison* group of
non-host units. The average minority percentage of the non-host zip code
areas in the UCC study was half that of the non-host census tracts in the
SADRI study. Thus, differences in the construction of the control
(comparison) populations used in the two respective studies, rather than
differences in the selection of their units of analysis, appeared to be the key
to understanding the differences in the two studies' findings.

    That the gap in the minority percentages between host and non-host areas
in the SADRI study was smaller than that of the UCC study suggested that
the areas excluded by SADRI had lower minority percentages. I obtained
a list from SADRI of the Standard Metropolitan Statistical Areas (SMSAs)
included in their study and found that this was indeed the case. Not only
were the minority percentages in the excluded metropolitan areas lower, but

among the demographic characteristics examined by SADRI, the African American percentage was the only one in which mean differences between metropolitan areas included vs. excluded by SADRI were statistically significant (Table 4). In a logistic regression, the African American percentage was furthermore the best predictor of whether or not a metropolitan area was included in the SADRI study (significance level of 0.0000; Table 5). Ironically, although the SADRI researchers had defended their approach of excluding from their study metropolitan areas not containing a hazardous waste Treatment, Storage, and Disposal Facility (TSDF) on the assumption that such areas likely do not serve "the markets that TSDFs currently serve" (Anderton et al., 1994, p. 246), I found that the mean values of their chief indicator of industrial activity – the proportion of people employed in precision manufacturing and labor occupations – to be nearly identical for included and excluded metropolitan areas (Table 4). Furthermore, it was the variable least able to predict which metropolitan areas contained a TSDF and which did not (significance level of 0.8935; Table 5).

In addition to the discrepant way in which the SADRI study constructed its control population, I was furthermore struck that when the SADRI researchers went on to examine the minority percentages of nearby tracts within 2.5 miles from the geographic centers (or "centroids") of the host tracts, the average minority percentage increased from 24% to 35%. When

Table 4. Comparison of Mean Demographics of SMSAs Containing a TSDF with Mean Demographics of SMSAs Not Containing a TSDF.

	Means for SMSAs Containing a TSDF (and Hence Included in SADRI Study)	Means for SMSAs Not Containing a TSDF (and Hence Excluded from SADRI Study)
% Black	11.55	7.79[****]
% Hispanic	5.69	5.36
% Families below poverty line	12.34	12.80
% Households receiving public assistance	7.50	7.01
% Males employed in civilian labor force	57.18	57.13
% Employed in precision occupations	31.16	31.03
Mean value of housing stock	54551.17	51449.15
N	152	166

[****]$p < 0.0000$. Statistical significance is based on t-tests.

Table 5.   Results of Logistic Regression Model Predicting Whether or
Not SMSA Contains a TSDF (and Hence Whether or Not SMSA Was
Included in SADRI Study).

Variable	Coefficient	Standard Error	Significance
% Black	0.0804	0.0167	0.0000
% Hispanic	0.0304	0.0140	0.0297
% Families below poverty line	−0.2321	0.0600	0.0001
% Households receiving public assistance	0.1729	0.0658	0.0086
% Males employed in civilian labor force	0.0487	0.0677	0.4724
% Employed in precision occupations	−0.0031	0.0234	0.8935
Mean value of housing stock	−0.0000	0.0000	0.8158
Constant	−1.9352	3.6733	0.5983
χ^2 (7 df)	37.559		0.0000

tracts within 2.5 miles of the host for the 25 largest metropolitan areas were
examined, the minority percentage increased again to 42% (Table 3). Based
on these outcomes, it seemed ironic to me that, rather than overestimating
the percentage of people of color around hazardous waste facilities, as the
SADRI researchers had implied, the UCC study may have actually
underestimated them. I wondered further what would happen to these
demographics if the 2.5-mile circular buffers were centered at the facilities
rather than at the host tract centroids.

REASSESSING RACIAL AND SOCIOECONOMIC
DISPARITIES IN ENVIRONMENTAL JUSTICE
RESEARCH

My work with the 1990 DAS, analysis of the locations of the hazardous
waste landfills in the Houston court case, and comparative analyses of
the UCC and SADRI studies led me to further contemplation about the
methodological issues involved in environmental justice research. The
results of these analyses made it increasingly clear that the unit-hazard
coincidence method – the method that was appearing to become the
standard approach in conducting environmental inequality analyses – was
problematic. Maps of hazardous waste facility locations in the Detroit
metropolitan area showed them often to be located near the boundaries of
their host tracts (e.g., see Facility A and other facilities in Fig. 1). Were non-
host tracts adjacent to these facilities also affected? Maps of the Detroit and

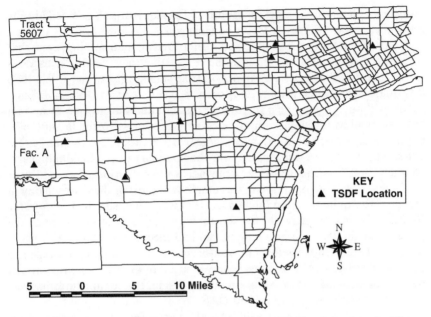

Fig. 1. 1990 Census Tracts and Hazardous Waste Facility Locations in Wayne County, Michigan.

Houston metropolitan areas clearly showed a great variation in the size and shape of the census tracts within them. Were large host tracts as effective in accounting for the proximity between hazardous waste facilities and nearby residential populations as small host tracts?

Through a series of grants, I was able to work with my then Ph.D. student and current co-author and colleague, Professor Robin Saha,[2] to map the precise geographic locations of all the hazardous waste treatment and storage facilities (TSDFs) in the U.S. We found that the pattern we observed in the Detroit metropolitan area extended elsewhere. That facilities are located near the boundaries of their host tracts is not a rare occurrence. Indeed, we found that, nationally, 49% of hazardous waste TSDFs are located within 0.25 miles of their host tract boundaries, while 71% are located within 0.5 miles (Mohai & Saha, 2006). If the environmental justice thesis is correct, that people of color and poor people are located disproportionately where environmental hazards are located, then would not the racial and socioeconomic characteristics of such adjacent tracts be more similar to the host tracts proper than non-host tracts much farther away (e.g., compare the tract just south of Facility A with Tract 5607 in the northwest corner of Wayne County, Michigan)?

We also found, as we had observed for the Detroit metropolitan area, that there is a great deal of variation in the size of the tracts hosting hazardous waste facilities. We found nationally that the smallest host tract is less than 0.1 square miles, while the largest is over 7500 square miles (Mohai & Saha, 2006). Although it may be a reasonable assumption that anyone living in a tract as small as 0.1 square miles necessarily lives close to the facility in it, the same cannot be said for the very large tract. We found, as in many other cases, that the facility was located near the boundary of the tract and nearly 35 miles away from its center or centroid. Rather than being located near the facility, a safer assumption in this case is that most of the people living in this tract live quite far from the facility. If most residents in the tract live far from the facility, is there any reason to expect an association between the presence of the facility and the racial and socioeconomic characteristics of the tract? It seemed to us that the unit-hazard coincidence method failed to control adequately for proximity, and we began to experiment with various ways of controlling for proximity by using distance-based methods, similar to what was employed in the 1990 DAS.

There was a catch. In the 1990 DAS, the respondents in the survey could be treated as geographic points and thus there was little ambiguity about whether or not they were located inside or outside specified distances to the hazardous sites. However, census units take up two-dimensional space, and when circular buffers of a specified distance from the hazardous sites are drawn, it is not always the case the units, such as tracts, lay completely inside or outside the buffers. Often, these units are only partially captured by the circular buffers. How should these partially captured units be assessed? We began to experiment with various approaches. At the same time, we discovered that other researchers had also begun to experiment with similar approaches, although most studies were limited in geographic scope and usually did not discuss the advantages of these approaches in controlling for proximity over the more widely used unit-hazard coincident method (see, e.g., Boer, Pastor, Sadd, & Snyder, 1997; Chakraborty & Armstrong, 1997; Glickman, 1994; Pastor, Sadd, & Hipp, 2001; Pollock & Vittas, 1995; Sheppard, Leitner, McMaster, & Tian, 1999).

Professor Saha and I outlined the various approaches in a recent article in the journal *Demography* (Mohai & Saha, 2006). One promising approach, modeled after that employed by the SADRI researchers (Anderton et al., 1994; Davidson & Anderton, 2000), is to include as part of the host neighborhood all tracts in which at least 50% of their areas are captured by circular buffers of specified distances from the hazards. We termed this the "50% areal containment method." A critical difference in our approach from that of the SADRI studies, however, is that we centered the radii at the facilities rather than at the centroids of the host tracts. We applied radii of various distances (up to three miles) around the nation's hazardous waste facilities to define the host neighborhoods (see Fig. 2a). An even more promising approach is the "areal apportionment" method (see, e.g., Chakraborty & Armstrong, 1997; Glickman, 1994; Hamilton & Viscusi, 1999; Sheppard et al., 1999). In applying this method, the populations of all census tracts wholly or partially captured by a radii of a specified distance are weighted by the proportion of the areas of the tracts that are captured, and the weighted populations are then summed or aggregated. This approach produces demographic estimates within perfectly circular host neighborhoods around the hazardous sites (see Fig. 2b). We found that the areal apportionment method produces especially reliable results. That is, regardless of whether zip code areas, census tracts, or block groups are used as the building block units, the estimated demographic characteristics within the circular neighborhoods produced by the areal apportionment method remain virtually constant (Mohai & Saha, 2007).

We demonstrated in our analyses that distance-based methods, such as the 50% areal containment and areal apportionment methods, better control for proximity between where hazardous sites and nearby residential proximity are located than does the more traditional unit-hazard coincidence method. We did this in two ways. First, we showed that distance-based methods using one-, two-, and three-mile radii define neighborhoods that are generally smaller than when only raw host tracts are used. For example, we found that the average census tract hosting a hazardous waste facility is 58.4 square miles, while a neighborhood around a hazardous waste facility defined by a one-mile radius using the 50% areal containment method is on average only 2.4 square miles (Mohai & Saha, 2006). Even at a distance of three miles from a facility, the average size of the neighborhood formed using 50% areal containment is less than half that of the average host census tract (21.8 vs. 58.4 square miles, respectively). Areal apportionment produces similar results to that of the 50% areal containment method, with neighborhoods defined by one- and three-mile radii containing average areas of 3.1 and 28.3 square miles, respectively.

Second, we showed that when distance-based methods are used, the hazard is more likely to stay close to the centroid (i.e., center) of the host neighborhood than when the unit-hazard coincidence method is employed. For example, we found that on average the locations of hazardous waste

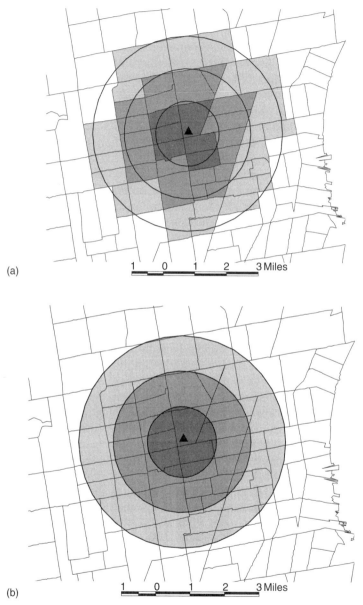

(a)

(b)

Fig. 2. Host Neighborhoods Surrounding Hazardous Waste Facility (a) Defined by 50% Areal Containment Using 1.0, 2.0, and 3.0 mile Radii. (b) Defined by Areal Apportionment Using 1.0, 2.0, and 3.0 mile Radii. Adapted from Mohai and Saha (2006).

facilities and the centroids of the neighborhoods defined by the 50% areal containment method, using either one- or three-mile radii, deviate less than one mile. In contrast, the average deviation between the location of hazardous waste facilities and the centroids of host census tracts is almost two miles. Furthermore, when the areal apportionment method is applied, there is no deviation between the location of the hazards and the centroids of the neighborhoods, regardless of the size of the radius. Because the areal apportionment method produces perfectly circular neighborhoods, hazardous sites are always located right at the center.

That the two distance-based methods generally produce smaller host neighborhoods (even when radii as large as three miles are used) than does the unit-hazard coincidence method, and that the hazards under investigation are located closer to the centroids of the respective neighborhoods, means that the distance-based methods do a better job of controlling for proximity between the location of the hazards and nearby residential populations. If we are able to define neighborhoods that are closer to the hazards of interest than is possible from using the unit hazard coincidence method, will we find that the proportions of people of color and poor people are greater near the hazards than what the unit-hazard coincidence method has previously revealed?

We indeed found that this was the case (Mohai & Saha, 2006, 2007). Using the 1990 census, we found the racial and socioeconomic disparities in the distribution of hazardous waste facilities to be much greater when either of the two distance-based methods were employed to define the neighborhoods than when the unit-hazard coincidence method was applied (see Table 6). For example, the difference in the nonwhite percentages between host and non-host tracts was only 1.2% using the unit-hazard coincidence method, but the difference in the nonwhite percentages of host and non-host neighborhoods defined by a one-mile radius using either the 50% areal containment or areal apportionment methods was over 20% (Table 6). Similarly, the difference in the poverty percentages between host and non-host tracts was only 0.5%, while the difference in the poverty percentages of host and non-host neighborhoods defined by a one-mile radius using either distance-based method was over 6.0%. Furthermore, racial and socioeconomic disparities between host and non-host neighborhoods defined by a *three*-mile radius using either distance-based method were also found to be greater than disparities found between host and non-host tracts (not shown in Table 6).

Given the better ability of distance-based methods over the unit-hazard coincidence method to control for proximity between hazardous sites and

Table 6. Comparisons of Population Characteristics in Host and Non-host Neighborhoods Defined by Areal Apportionment and 50% Areal Containment Methods vs. Unit-hazard Coincidence Method.

	Unit-hazard Coincidence		Areal Apportionment Using One-mile Radius		50% Areal Containment Using One-mile Radius	
	All Host Tracts	All Non-host Tracts	Within One mile	Beyond One mile	Within One mile	Beyond One mile
% African American	12.7	12	18.8	11.9	20.2	11.9
% Hispanic	10.1	8.8	20.1	8.6	21.8	8.7
% Nonwhite	25.4	24.2	42.8	24	46.2	24
% Below poverty line	13.6	13.1	19.1	13	20.6	13
Mean household income	$34,526	$38,491	$31,977	$38,545	$30,598	$38,543
Mean property value	$88,892	$111,883	$92,442	$110,320	$89,747	$111,856
% Without high school diploma	28.5	24.7	34.4	24.6	36.5	24.6
% With college degree	14.4	20.4	14.1	20.4	13.1	20.4
% Employed in executive, management, or professional occupation	21.4	26.4	20.7	26.4	19.7	26.4
% Employed in precision production, transportation, or labor occupation	31.4	26.1	31.3	26.1	32.2	26.1
% Unemployed	6.7	6.3	8.9	6.3	9.6	6.3

Note: Demographic values are based on the 1990 census.
Source: Adapted from Mohai and Saha (2007).

nearby residential populations, and the potential for the unit-hazard coincidence method to produce misleading results, its continued use for conducting environmental justice analyses seems difficult to justify. This seems especially so given the current wide availability of Geographic Information System (GIS) technology. I anticipate that the application of distance-based methods will reveal far greater racial and socioeconomic disparities than previously reported in the distribution of environmental hazards of a wide variety.

THE RACE VS. CLASS DEBATE

There has been much interest in environmental justice research to determine whether racial disparities in the distribution of environmental hazards are independent of socioeconomic disparities. There are several reasons why such a determination is of interest. From an academic standpoint, knowing whether or not racial disparities in the distribution of environmental hazards are independent of socioeconomic disparities is an important step in narrowing down explanations of why the environmental disparities exist. If in a multivariate statistical analysis controlling for the average incomes or average property values in a community eliminates the significance of the race variables (e.g., the African American or Hispanic percentages of the community) in predicting the presence of environmental hazards, this suggests economic explanations of the distribution of hazardous waste and industrial sites may be more salient than racial explanations. For example, such facilities may tend to be located where racial and ethnic minorities live because industries desire to minimize production costs by siting new facilities in places where land values are low (Boone and Madarres, 1999; Hamilton, 1995; Hird & Reese, 1998; Rhodes, 2003). These may be coincidentally where minorities live. Alternatively, the facilities, once sited, may cause a decline in property values and quality of life, motivating affluent whites to move away and the poor and minorities to move in because of increased affordability of housing. However, if race variables remain significant after controlling for economic variables, it may suggest that factors uniquely related to race play a role. For example, housing discrimination may limit the ability of people of color to move away from such hazardous sites, beyond the limitations imposed by constrained incomes (Mohai & Bryant, 1992; Szasz & Meuser, 2000). Or perhaps minority communities are targeted for new facilities because minority communities over time have come to be recognized as the "paths of least resistance" by government and industry (Bullard, 1990; Bullard & Wright, 1987b; Hurley, 1995; Pellow, 2002; Pulido, Sidawi, & Vos, 1996). The debate about whether or not race plays an independent role in the distribution of environmental hazards also has political and policy implications. The first draft of the U.S. EPA's report, *Environmental Equity: Reducing Risks for All Communities*, made no mention of race in the distribution of such hazards, focusing instead on only the socioeconomic characteristics of overburdened communities. Subsequently, an internal EPA memo was leaked to the public revealing EPA's concerns about "the minority fairness issue reach[ing] a flashpoint – that stage in an emotionally charged public controversy when activist groups finally succeed in persuading the more influential mainstream groups (civil rights organizations, unions, churches) to take ill-advised actions" (Lester, Allen, & Hill, 2001, p. 46). Clearly, there were concerns among some in the EPA about the potency of this issue framed in racial terms. In response to the criticisms of environmental justice activists and scholars, the EPA acknowledged the racial dimensions of environmental inequalities in its final report. As a practical policy outcome, by recognizing the racial aspects of environmental disparities, it allowed the agency to begin to shape and implement environmental justice policy by drawing on civil rights laws, principally Title VI of the 1964 Civil Rights Act. Recent Supreme Court rulings, however, have seriously limited EPA's and other agencies' abilities to use existing civil rights laws to enforce environmental justice policies.

Supporting environmental justice activists' claims, systematic reviews of the quantitative research have tended to find that race effects are stronger than economic effects (Mohai & Bryant, 1992; Goldman, 1994; Ringquist, 2005). However, there is nevertheless a range in the magnitude of racial and socioeconomic disparities found, with some studies finding none. In a recent meta-analysis of 49 quantitative studies, Ringquist (2005) found that race effects are stronger than economic effects but cautioned that these are nevertheless modest. An argument we made in our article in *Demography* (Mohai & Saha, 2006), however, is that if the unit-hazard coincidence method does a poor job of controlling for proximity, and hence leads to an underestimation of the magnitude of both racial and socioeconomic disparities in the distribution of environmentally hazardous sites, then the results of multivariate analyses involving the unit-hazard coincidence method aimed at assessing the relative importance of race and socio-economic variables will also be unreliable. We demonstrated that when distance-based methods are employed, the outcomes of multivariate analyses are different, and hence assessments about the relative importance of race

Table 7. Logistic Regression Results Comparing Unit-hazard
Coincidence and 50% Areal Containment Methods.

(1) Variables	Unit-hazardous Coincidence		50% Areal Containment (One-mile Radius)	
	(2) Coefficient	(3) Significance	(4) Coefficient	(5) Significance
% African American	−0.003	0.986	0.698	0.000
% Hispanic	0.431	0.066	1.482	0.000
Mean hold income ($1000s)	0.012	0.000	−0.025	0.000
Mean property value ($1000s)	−0.002	0.058	0.005	0.000
% With college degree	0.338	0.673	−1.704	0.012
% Employed in executive, management, or professional occupation	−3.215	0.002	−0.872	0.282
% Employed in precision production, transportation, or labor occupation	2.323	0.000	1.787	0.000
Constant	−5.052	0.000	−4.197	0.000
−2 Log likelihood	6010.2		8077.3	
Model χ^2	153.743	0.000	548.233	0.000
Sample size	59,050		59,050	

Source: Adapted from Mohai and Saha (2006).

and socioeconomic variables (and hence racial and socioeconomic explanations of environmental inequality) are changed (Mohai & Saha, 2006).

Results of the logistic regression analyses in Table 7 illustrate this point. In the analysis involving the unit-hazard coincidence method, the dependent variable takes a value of one if a tract hosts a hazardous waste facility and a value of zero if it does not. In the analysis involving one of the distance-based methods (specifically, 50% areal containment), the dependent variable takes a value of one if a tract is within one mile of a hazardous waste facility and a value of zero if it is not. Results for the two regression analyses are dramatically different. In the logistic regression involving the unit-hazard coincidence method, neither the African American percentage nor the Hispanic percentage are statistically significant (at the 0.05 level) predictors of hazardous waste facility location, while the occupation variables are. Specifically, the larger the proportion of residents employed in manufacturing and labor jobs, and the smaller the proportion employed in executive, management, and professional jobs, the greater the likelihood there will be a hazardous waste facility nearby. Such an outcome might be interpreted to mean that any racial disparities found in the bivariate analyses are explained by the disproportionate number of African Americans in manufacturing jobs vs. professional and management jobs, and that hazardous waste facilities tend to be located where manufacturing labor pools are nearby. In contrast, in the logistic regression associated with the 50% areal containment method the race variables remain highly statistically significant, indicating that the racial disparities are not purely a function of the occupational and other socioeconomic variables. Some other factors associated with race appear to account for the disparities. In this case, housing discrimination or the possible targeting of minority communities by government and industry have not been ruled out. Whereas the results of the first regression motivate further exploration of economic explanations of environmental inequality and not racial ones, the results of the second regression clearly motivate further exploration of the latter.

CURRENT AND FUTURE RESEARCH

Better methods in environmental justice research should lead to more accurate assessments about the magnitude of racial and socioeconomic disparities in the distribution of environmental hazards as well as to more accurate hypothesis tests regarding cause and effect relationships in hazard distribution. However, most quantitative research to date has been based on analyzing present-day disparities from cross-sectional data, and to determine how present-day disparities have come about requires analysis of longitudinal data. This is because present-day disparities may have resulted from two processes: (a) disproportionate siting of hazardous waste facilities and other locally unwanted land uses in people of color communities *at the time* of siting or (b) demographic changes *after* facility siting. Of course, both processes working together could account for present-day disparities. "Which came first, people or pollution?" is the chicken or egg question in environmental justice research.

An answer to this question is important because it can help researchers to understand the social, economic, and political processes by which environmental disparities occur. It can also help in crafting policies to prevent environmental disparities from occurring. For example, if present-day environmental disparities occur largely because of a pattern of disproportionately placing facilities in poor and people of color communities at the time of siting, then policies that focus on managing the siting and permitting process to avoid such disparities could be adopted. If, however, the present-day disparities occur because of demographic changes after siting, then policies managing the siting process may not be enough. Policies that ensure newcomers adequate access to information about the potential health risks in a neighborhood (so that they can weigh the risks against the desirability and affordability of the housing) and policies that make certain that people of color are not intentionally steered into contaminated neighborhoods should be implemented (Burby & Strong, 1997; Pastor et al., 2001). Currently, there are very few longitudinal studies providing an answer to the "chicken or egg" question, and those that do exist provide contradictory findings.

That there are currently few longitudinal studies is partly the result of the difficulty of doing them. First, researchers need to get reliable information about when facilities were sited so that they can check census data *at or near* the time of siting, not the most recent census, to see if there has been a pattern of disproportionately placing the facilities in people of color communities. For example, if a facility was sited in 1970, one needs to examine the demographics around 1970, not 2000, since 2000 census data may reflect demographic changes around the site since 1970. A second major difficulty in conducting longitudinal studies is the fact that census tract boundaries can shift from decade to decade. This makes it difficult to know whether demographic changes in an area are the result of population shifts or simply tract boundary shifts. Finally, longitudinal studies employing the unit-hazard coincidence method are susceptible to the same limitations as cross-sectional studies concerning adequate control of proximity between sites and nearby residential populations. Because of these difficulties, the number of long-itudinal studies attempting to answer the chicken or egg question are not only few, but those that have been done have led to contradictory findings.

For example, longitudinal studies using the unit-hazard coincidence method have been inconclusive as to whether disproportionate siting or post-siting demographic change explains present-day disparities (Been & Gupta, 1997; Oakes et al., 1996). In contrast, studies employing distance-based methods have provided clear evidence of patterns of racially disproportionate exposure at the time of siting (Pastor et al., 2001; Saha & Mohai, 2005). However, although the patterns from these latter studies are clear, they are based on limited geographic samples (e.g., the Pastor et al. study is focused on the Los Angeles metropolitan area, while the Saha and Mohai study is focused on the state of Michigan). Thus, one of the projects currently being worked on by Professor Saha and me is to conduct a national-level longitudinal analysis of the demographics around commercial hazardous waste facilities using distance-based methods. Distance-based methods not only better control for proximity, but they also provide a means of stabilizing over time geographic boundary changes around the facilities. In addition to more work being needed to answer the environmental justice "chicken or egg" question and test hypotheses about the causes of present-day environmental disparities, more work is also needed to model pollution exposures and assess the role such exposures play in accounting for racial and socioeconomic disparities in health and mortality. I am working with colleagues at the University of Michigan and elsewhere around the country to help advance these projects.

NOTES

1. Empirical evidence has tended to show that African Americans are as concerned as white Americans about a wide range of environmental issues, including nature preservation and global environmental issues. African Americans are even more concerned than white Americans about environmental issues that have implications for human health, such as hazardous waste and pollution issues. Mohai and Bryant (1998) have demonstrated that this is related to the disproportionate burden of environmental hazards in African American neighborhoods.
2. Robin Saha is currently Assistant Professor of Environmental Studies at the University of Montana, and is also affiliated with its School of Public and Community Health Sciences.

ACKNOWLEDGEMENTS

I wish to thank the following environmental sociologists for their inspiration and mentorship over the years: Richard Bord, Steven R. Brechin, Robert D. Bullard, Frederick H. Buttel (deceased), Riley Dunlap, Louise Fortmann, William Freudenburg, Craig Humphrey, James Kennedy, Richard Krannich, Dorceta E. Taylor, Ben W. Twight, Edward Walsh, Patrick C. West, and Beverly Wright. I am especially indebted to Bunyan Bryant and Robin Saha for their friendship and our long and productive collaborations together.

REFERENCES

Anderton, D. L., Anderson, A. B., Oakes, J. M., & Fraser, M. R. (1994). Environmental equity: The demographics of dumping. *Demography, 31*, 229–248.

Anderton, D. L., Oakes, J. M., & Egan, K. L. (1997). Environmental equity in superfund: Demographics of the discovery and prioritization of abandoned toxic sites. *Evaluation Review, 21*, 3–26.

Asch, P., & Seneca, J. J. (1978). Some evidence on the distribution of air quality. *Land Economics, 54*, 278–297.

Been, V., & Gupta, F. (1997). Coming to the nuisance or going to the barrios? A longitudinal analysis of environmental justice claims. *Ecology Law Quarterly, 24*, 1–56.

Berry, B. J. L., Caris, S., Gaskill, D., Kaplan, C. P., Piccinini, J., Planert, N., Rendall, J. H., III., & de Ste. Phalle, A. (1977). *The social burdens of environmental pollution: A comparative metropolitan data source.* Cambridge, MA: Ballinger Publishing Co.

Boer, J. T., Pastor, M., Sadd, J. L., & Snyder, L. D. (1997). Is there environmental racism? The demographics of hazardous waste in Los Angeles. *Social Science Quarterly, 78*, 793–810.

Boone, C. G., & Madarres, A. (1999). Creating a toxic neighborhood in Los Angeles County: A historical examination of environmental inequality. *Urban Affairs Review, 35*, 163–187.

Bryant, B., & Mohai, P. (Eds). (1992a). *Race and the incidence of environmental hazards: A time for discourse.* Boulder, CO: Westview Press.

Bryant, B., & Mohai, P. (1992b). The Michigan conference: A turning point. *EPA Journal, 18*, 9–10.

Bullard, R. D. (1983). Solid waste sites and the Houston black community. *Sociological Inquiry, 53*, 273–288.

Bullard, R. D. (1990). *Dumping in dixie: Race, class, and environmental quality.* Boulder, CO: Westview Press.

Bullard, R. D., & Wright, B. H. (1986). The politics of pollution: Implications for the black community. *Phylon, 47*, 71–78.

Bullard, R. D., & Wright, B. H. (1987a). Blacks and the environment. *Humboldt Journal of Social Relations, 14*, 165–184.

Bullard, R. D., & Wright, B. H. (1987b). Environmentalism and the politics of equity: Emergent trends in the black community. *Mid-American Review of Sociology, 12*, 21–38.

Burby, R. J., & Strong, D. E. (1997). Coping with chemicals: Blacks, white, planners, and industrial pollution. *Journal of the American Planning Association, 63*, 469–480.

Burch, W. R. (1976). The peregrine falcon and the urban poor: Some sociological interrelations. In: P. Richerson & J. McEvoy (Eds), *Human ecology, an environmental approach* (pp. 308–316). Belmont, CA: Duxbury Press.

Buttel, F. H., & Flinn, W. L. (1978). Social class and mass environmental beliefs: A reconsideration. *Environment and Behavior, 10*, 433–450.

Chakraborty, J., & Armstrong, M. P. (1997). Exploring the use of buffer analysis for the identification of impacted areas in environmental equity assessment. *Cartography and Geographic Information Systems, 24*, 145–157.

Council on Environmental Quality (CEQ). (1971). *The second annual report of the council on environmental quality.* Washington DC: U. S. Government Printing Office.

Davidson, P., & Anderton, D. L. (2000). The demographics of dumping II: Survey of the distribution of hazardous materials handlers. *Demography, 37*, 461–466.

Freeman, A. M. (1972). The distribution of environmental quality. In: A. V. Kneese & B. T. Bower (Eds), *Environmental quality analysis.* Baltimore, MD: Johns Hopkins University Press for Resources for the Future.

Gelobter, M. (1988). The distribution of air pollution by income and race. Paper presented at the *2nd symposium on social science in resource management*, Urbana, Illinois, June.

Gelobter, M. (1992). Toward a model of environmental discrimination. In: B. Bryant & P. Mohai (Eds), *Race and the incidence of environmental hazards: A time for discourse* (pp. 64–81). Boulder, CO: Westview Press.

Gianessi, L., Peskin, H. M., & Wolff, E. (1979). The distributional effects of uniform air pollution policy in the U.S.. *Quarterly Journal of Economics, 93*, 281–301.

Glickman, T. (1994). Measuring environmental equity with GIS. *Renewable Resources Journal, 12*, 17–21.

Goldman, B. A. (1994). *Not just prosperity: Achieving sustainability with environmental justice.* Washington DC: National Wildlife Federation.

Hamilton, J. T. (1995). Testing for environmental racism: Prejudice, profits, political power? *Journal of Policy Analysis and Management, 14*, 107–132.

Hamilton, J. T., & Viscusi, K. W. (1999). *Calculating risks? The spatial and political dimensions of hazardous waste policy.* Cambridge, MA: MIT Press.

Handy, F. (1977). Income and air quality in Hamilton, Ontario. *Alternatives, 6*, 18–24.

Harrison, D., Jr.. (1975). *Who pays for clean air: The cost and benefit distribution of automobile emission standards.* Cambridge, MA: Ballinger.

Hird, J. A., & Reese, M. (1998). The distribution of environmental quality: An empirical analysis. *Social Science Quarterly, 79*, 694–716.

Hurley, A. (1995). *Environmental inequities: Class, race and industrial pollution in gary, Indiana, 1945–1990.* Chapel Hill, NC: University of North Carolina Press.

Kruvant, W. J. (1975). People, energy, and pollution. In: D. K. Newman & D. Day (Eds), *The American energy consumer* (pp. 125–167). Cambridge, MA: Ballinger.

Lester, J. P., Allen, D. W., & Hill, K. M. (2001). *Environmental injustice in the United States: Myths and realities.* Boulder, CO: Westview Press.

McMaster, R. B., Leitner, H., & Sheppard, E. (1997). GIS-based environmental equity and risk assessment: Methodological problems and prospects. *Cartography and Geographic Information Systems, 24*, 172–189.

Mitchell, R. C. (1979). Silent springs/solid majorities. *Public Opinion, 2*, 16–2055.

Mohai, P. (1985). Public concern and elite involvement in environmental-conservation issues. *Social Science Quarterly, 66*, 820–838.

Mohai, P. (1990). Black environmentalism. *Social Science Quarterly, 71*, 744–765.

Mohai, P. (1995). The demographics of dumping revisited: Examining the impact of alternate methodologies in environmental justice research. *Virginia Environmental Law Journal, 13*, 615–653.

Mohai, P. (2003). Dispelling old myths: African American concern for the environment. *Environment, 45*, 10–26.

Mohai, P., & Bryant, B. (1992). Environmental racism: reviewing the evidence. In: B. Bryant & P. Mohai (Eds), *Race and the incidence of environmental hazards: A time for discourse* (pp. 163–176). Boulder, CO: Westview Press.

Mohai, P., & Bryant, B. (1998). Is there a "race" effect on concern for environmental quality? *Public Opinion Quarterly, 62*, 475–505.

Mohai, P., & Kershner, D. (2002). Race and environmental voting in the U.S. Congress. *Social Science Quarterly, 83*, 167–189.

Mohai, P., & Saha, R. (2006). Reassessing racial and socioeconomic disparities in environmental justice research. *Demography, 43*, 383–399.

Mohai, P., & Saha, R. (2007). Racial inequality in the distribution of hazardous waste: A national-level reassessment. *Social Problems, 54*, 343–370.

Oakes, J. M., Anderton, D. L., & Anderson, A. B. (1996). A longitudinal analysis of environmental equity in communities with hazardous waste facilities. *Social Science Research, 25,* 125–148.

Pastor, M., Sadd, J., & Hipp, J. (2001). Which came first? Toxic facilities, minority move-in, and environmental justice. *Journal of Urban Affairs, 23,* 1–21.

Pellow, D. N. (2002). *Garbage wars: The struggle for environmental justice in Chicago.* Cambridge, MA: MIT Press.

Pollock, P. H., & Vittas, E. M. (1995). Who bears the burdens of environmental pollution? Race, ethnicity, and environmental equity in Florida. *Social Science Quarterly, 76,* 294–310.

Pulido, L., Sidawi, S., & Vos, R. O. (1996). An archaeology of environmental racism in Los Angeles. *Urban Geography, 17,* 419–439.

Rhodes, E. L. (2003). *Environmental justice in America: A new paradigm.* Bloomington, IN: Indiana University Press.

Ringquist, E. J. (2005). Assessing evidence of environmental inequities: A meta-analysis. *Journal of Policy Analysis and Management, 24,* 223–247.

Saha, R., & Mohai, P. (2005). Historical context and hazardous waste facility siting: Understanding temporal patterns in Michigan. *Social Problems, 52,* 618–648.

Sheppard, E., Leitner, H., McMaster, R. B., & Tian, H. (1999). GIS-based measures of environmental equity: Exploring their sensitivity and significance. *Journal of Exposure Analysis and Environmental Epidemiology, 9,* 18–28.

Szasz, A., & Meuser, M. (2000). Unintended, inexorable: The production of environmental inequalities in Santa Clara County, California. *American Behavioral Scientist, 43,* 602–632.

Taylor, D. E. (1989). Blacks and the environment: Toward explanation of the concern and action gap between blacks and whites. *Environment and Behavior, 21,* 175–205.

United Church of Christ (UCC). (1987). *Toxic wastes and race in the United States: A national report on the racial and socioeconomic characteristics of communities with hazardous waste sites.* New York, NY: Commission for Racial Justice, United Church of Christ.

U.S. Environmental Protection Agency (EPA) (1992). *Environmental equity: Reducing risks for all communities.* EPA 230-R-92-008, June. Washington DC: U.S. Environmental Protection Agency.

U.S. General Accounting Office (GAO). (1983). *Siting of hazardous waste landfills and their correlation with racial and economic status of surrounding communities.* Washington DC: U.S. General Accounting Office.

Van Liere, K. D., & Dunlap, R. E. (1980). The social bases of environmental concern: A review of hypotheses, explanations and empirical evidence. *Public Opinion Quarterly, 44,* 181–197.

West, P. C., Mark Fly, J., Larkin, F., & Marans, R. (1992). Minority anglers and toxic fish consumption: Evidence from a state-wide survey of Michigan. In: B. Bryant & P. Mohai (Eds), *Race and the incidence of environmental hazards: A time for discourse* (pp. 100–113). Boulder, CO: Westview Press.

Zupan, J. M. (1973). *The distribution of air quality in the New York region.* Johns Hopkins.

EQUITY, UNNATURAL MAN-MADE DISASTERS, AND RACE: WHY ENVIRONMENTAL JUSTICE MATTERS ☆

Robert D. Bullard

ABSTRACT

This chapter chronicles some of the early years of the author growing up in the racially segregated South Alabama and its influence on his thinking about race, environment, social equity, and government responsibility and his journey to becoming an environmental sociologist, scholar, and activist. Using an environmental justice paradigm, he uncovers the underlying assumptions that contribute to and produce unequal protection. The environmental justice paradigm provides a useful framework for examining and explaining the spatial relation between the health of marginalized populations and their built and natural environment, and government response to natural and man-made disasters in African American communities. Clearly, people of color communities have borne a disproportionate burden and have received differential treatment from government in its response to health threats such as childhood lead

☆Paper submitted to Research in Social Problems and Public Policy (RSPPP) Special issue on Equity and the Environment.

Equity and the Environment
Research in Social Problems and Public Policy, Volume 15, 51–85
ISSN: 0196-1152/doi:10.1016/S0196-1152(07)15002-X

poisoning, toxic waste and contamination, industrial accidents, hurricanes, floods and related weather-related disasters, and a host of other man-made disasters. The chapter brings to the surface the ethical and political questions of "who gets what, why, and how much" and why some communities get left behind before and after disasters strike.

The environmental justice paradigm provides a useful framework for examining and explaining the spatial relation between the health of marginalized populations and their built and natural environment, and government response to disasters that affect those marginalized communities. More than three decades of environmental justice research clearly document that people of color and their communities have borne a disproportionate burden of pollution from incinerators, smelters, sewage treatment plants, chemical plants, industrial accidents, and a host of other man-made disasters (Bullard, 2005a).

The environmental justice framework incorporates other social movements and principles that seek to prevent and eliminate harmful practices in land use, industrial planning, health care, waste disposal, and sanitation services. The environmental justice framework attempts to uncover the underlying assumptions that may contribute to and produce unequal protection. This framework brings to the surface the ethical and political questions of "who gets what, why, and how much" (Bullard, 2005a, 2005b).

For me, environmental justice embraces the principle that all communities have a right to equal protection of our environmental, health, housing, employment, transportation, and civil right laws and regulations (Bullard, 2000). I used this definition years before the federal Environmental Protection Agency (EPA) decided environmental justice was worthy of a definition. The U.S. EPA defines environmental justice as the "fair treatment and meaningful involvement of all people regardless of race, color, national origin, or income with respect to the development, implementation, and enforcement of environmental laws, regulations and policies. Fair treatment means that no group of people, including racial, ethnic, or socio-economic groups, should bear a disproportionate share of the negative environmental consequences resulting from industrial, municipal, and commercial operations or the execution of federal, state, local, and tribal programs and policies" (U.S. EPA, 1998, p. 1).

In response to growing public concern and mounting scientific evidence, President Clinton on February 11, 1994 (the second day of a national health symposium that will be noted below) issued Executive Order 12898, "Federal Actions to Address Environmental Justice in Minority

Populations and Low-Income Populations" (Clinton, 1994). This Order attempts to address environmental injustice within existing federal laws and regulations (Council on Environmental Quality, 1997).

This chapter chronicles some of my early years as a youth growing up in the racially segregated South Alabama and their influence on my thinking about race, environment, social equity, and government responsibility. It also traces my journey to becoming an environmental sociologist, scholar, and activist. You should know up front that my hero is W.E.B DuBois, who founded the sociology department at Atlanta University (now Clark Atlanta University since the merger with Clark College in 1988), where I have an endowed chair.

Over the span of two decades, 1987–2007, I have written more than a dozen books that address environmental justice, environmental racism, land use and industrial facility siting, transportation, suburban sprawl, smart growth, livable communities, regional equity, and sustainable development. All of the books use an environmental justice framing.

MAKING THE RACE-ENVIRONMENT CONNECTION: MY EARLY ELBA YEARS

Much of my environmental justice writings over the past quarter century are rooted in my early years growing up in the South – the region that gave rise to the modern civil rights movement in Montgomery, Alabama in December 1955 – with Rosa Park's single defiant act of refusing to give up her seat on the bus to a white man. I was in the third grade at Elba (Alabama) Colored Elementary School when the Montgomery Bus Boycott began. I grew up just 83 miles south of Montgomery.

Elba is located in Coffee County, Alabama. The 2000 population was 4,185. The racial makeup of the Elba is 64 percent white, 34 percent black, and 2 percent other races. Elba is a typical small southern town, where blacks and whites live in segregated neighborhoods. Although blacks comprise one-third of the city's population, whites "run" the town as if its African American citizens were invisible. As late as 1964, a decade after *Brown v. Board of Education*, Elba's blacks attended segregated schools and were still being served at the back door of restaurants or served at the "colored" window (even though the "colored" signs had come down by then). I graduated from all-black Mulberry Heights High School in 1964. The segregated signs had come down in 1954, but life was still "separate and unequal" for blacks and whites in most places in Alabama.

Elba's white business and political elites used their white privileges to maintain their "edge," including access to the best land and safest physical environment. In Elba, Jim Crow segregation translated into white neighborhoods receiving libraries, street lighting, paved roads, sewer and water lines, garbage pick-up, swimming pools, and flood control measures, years before black neighborhoods received these tax-supported services. All-white mayors and city councils decided what was "best" for their town. That usually translated into blacks getting less than their fair share of the "goods" and more than their fair share of the "bads."

Elba city government systematically displaced black families from their downtown neighborhoods, where they had lived for decades, using eminent domain. City fathers displaced my grandmother in the 1960s from her near-town home. My grandmother later built a brick house next to my parents' home in Elba's all-black Mulberry Heights neighborhood.

The black landowners' property, including my grandmother's land, was subsequently sold to white families to build new homes and businesses in an area protected by the city's intricate levee system. Many displaced black families settled in the mostly black Mulberry Heights neighborhood, an area that frequently flooded each year because of poor drainage and rudimentary flood control measures. Storm runoff was channeled through poorly maintained, open ditches that often flooded the neighborhoods after heavy rains. For decades, Elba's segregated black neighborhoods flooded while white neighborhood remained high and dry. I recall on numerous occasions wading through ankle-deep water to get to school.

Because of its precarious location (the town sits at the confluence of Beaver Dam Creek, Whitewater Creek, and the Pea River), Elba has repeatedly flooded over the years. Even its sophisticated levee system has not been able to hold back "Mother Nature's" floodwaters. The first of four floods that occurred during the 20th century in Elba came in March 1929, the inaugural year of the Great Depression. That flood resulted in the construction of a levee in 1930, built as part of President Franklin D. Roosevelt's New Deal, that surrounds the entire perimeter of the downtown area.

Elba residents were forced to take refuge on housetops as they awaited rescue from rapidly rising floodwaters. Rains beginning in late February resulted in flooding that left 15,000 south Alabamians homeless. Although the Flood of 1929 hit Elba the hardest, several other Alabama towns, including Geneva and Brewton, were covered in as much as 15 feet of water.[1] I grew up hearing stories from my grandmother of the "Great Elba Flood of 1929" in which she lost a brother.

Despite the attempts to hold back the waters, Elba experienced severe flooding and levee breaches again in 1990, 1994, and 1998.[2] Although I was long gone from the town in the 1990s, many of my relatives were stranded and had to be evacuated. Elba's flooding problem stems from failure of the town's protective ring levee, which causes major flooding throughout the downtown area – as happened in 1990 and 1998 – and storm water accumulation within the levee, in the low-lying south-central and south-western areas of town, as happened in 1994.

After the 1994 flood, Elba applied for a hazard mitigation grant from FEMA to install a storm water drainage system. FEMA approved the grant application in July of that year. The system was built in 1997 by widening an existing drainage channel and installing two pumps at low-lying points in the town's southeast quarter. The pumps, designed to remove water quickly from flooded areas, are each capable of moving 17,500 gallons per minute (Federal Emergency Management Agency, 1999).

In 1998, Beaver Dam Creek eroded through the levee south of Elba, resulting in 2 feet of water flooding the town in minutes, eventually rising to 6 feet deep. Two people died in an automobile that was swept away. As a result of the 1994 and 1998 floods, the President issued a long-term Recovery Action Plan that included upgrading of the levee in April 1998.

On May 13, 2002, the groundbreaking ceremony for the rebuilding of the levee around the town of Elba (Alabama) was conducted. Photographs from the ceremony tell the story: Not a single African American attended the groundbreaking ceremony,[3] – not a small point in a town where one of every three residents is black. The total cost of the project is estimated to be $12,900,000. Nonfederal cost share paid by the State of Alabama will be $4,655,000. For comparison purposes, the cost of the 1990 flood was $150,000,000, including clean up, restoration, and the relocation of the schools to higher ground.

Elba has rebuilt after each major flood. It has resisted relocation. A 2004 article entitled "Elba Endures Rough Waters" in *The Crimson White* – the student-run newspaper of the University of Alabama in Tuscaloosa – describes the repeated experiences with flooding and the political leaders' resistance to relocating their hometown. Some government officials suggested that Elba should be moved to an alternate location on the U.S. 84. Local leaders at several town meetings soon dismissed this idea. Nobody knows how long the new levees, completed in 2004 by the Army Corps of Engineers, will hold. Nevertheless, the city leaders have garnered enough political and economic support to rebuild the town at its current site.

Elba's rebuilding effort has been spotty at best. In October 2005, some downtown buildings were still not occupied. I visited my hometown in August 2005 to take photographs for a report on government response to emergencies in African American communities that my colleague Beverly Wright and I had begun some nine months before Hurricane Katrina and floodwaters drowned New Orleans.[4]

The 1998 flood was not kind to Elba's downtown. I observed boarded-up stores and unoccupied buildings. I also observed hard-hit residences along Taylor Mill Road, a major thoroughfare through the mostly black Mulberry Heights community, that have not recovered from the 1998 flood. Elba's old "city dump" was located on Taylor Mill Road when I was a youth.

Not having a park or a swimming pool for blacks, Elba's old open burning dump and nearby Beaver Dam Creek were exciting places to play for a group of black boys. I also remember playing at open city dump on Harden Road in Opp, Alabama – where my maternal grandmother lived. The dump was later closed and public housing built on the site. My friends and I had no idea about the potential danger and potential health effects. We just knew it was a fun place. We went there even though our parents and grandparents scolded us for doing so.

The same held true for swimming in dangerous waters. My parents punished me on numerous occasions for "going down to the river." I swam many days in Elba's rivers and creeks. Most black boys my age in Elba who learned how to swim, learned in the Beaver Dam Creek, White Water River, or Pea River. And every so often a black kid would drown. My younger brother, who is eight years my junior, learned to swim in a segregated swimming pool built in Mulberry Heights much later – next to the school. On my recent visit to the old neighborhood, I looked for the swimming pool – it was no longer there. My old school was also boarded up and abandoned.

Elba's Mulberry Height neighborhood has experienced systematic decline and neglect of homes, streets, and overall quality of life, and it appeared to be a long way from rebounding from the 1998 flood. It was rather depressing to see what had happened to the neighborhood where I had spent the first 18 years of my life.

The hard-hit black neighborhoods have not recovered from the 1990, 1994, or 1998 floods. Many black inhabitants have sunk deeper into poverty after each flood. Seven years after the last major flood, many streets in Elba's black neighborhoods are not maintained, and scores of homes are boarded up. Few opportunities exist for blacks in this small southern town. Each subsequent flood worsened their chance of economic mobility. Generally, high school graduation was a one-way ticket out of town.

I left Elba in 1964 to attend college. I would visit periodically to visit my parents and relatives. In April 1968, when Dr. Martin Luther King, Jr. was assassinated in Memphis, Tennessee – over environmental and economic justice for striking black garbage workers – I was a senior at Alabama A&M University, a historically black university located in Huntsville, Alabama. After graduating from college in 1968, I wanted no part of Alabama and took a teaching job at Beaumont High School in St. Louis, Missouri. I taught high school from August to November – and was drafted in December 1968. I spent two years in the U.S. Marine Corps – from 1968–1970 – during the height of the Vietnam War.

IMPACT OF MY HOUSTON YEARS

After graduating from Atlanta University in 1972, I attended Iowa State University, where I received my Ph.D. in sociology in 1976. I moved to Houston in 1976 to take a job at historically black Texas Southern University. At the time, Houston was tagged the "golden buckle of the Sunbelt" and the "petro-chemical capital of the world." A decade after Dr. King was killed over racism and garbage, I would find myself drafted into a garbage struggle in Houston, Texas. In 1978, just two years out of graduate school, my attorney wife, Linda McKeever Bullard, asked me to conduct research for a lawsuit she had filed seeking to block the second largest waste disposal company in the world from locating a "sanitary landfill" (we all know there is nothing "sanitary" about a place where household garbage is dumped) in the midst of a predominately black Houston neighborhood.

My previous research on housing discrimination and residential segregation proved useful in making the transition to examining the location of landfills, incinerators, and garbage dumps and the racial composition of the neighborhoods. The central theme of my analysis was that all communities are not created equal when it comes to the siting of locally unwanted land uses (LULUs) such as garbage dumps, landfills, and incinerators. Instead, race and class dynamics, along with political disenfranchisement, interact to place some communities at special environmental and health risk from waste facility siting.

Using 10 graduate students in my research methods class, I embarked on a mission to track the location of solid waste facility siting in Houston from the beginning of the city's formation up to 1978 – when the permit for the Whispering Pine Sanitary Landfill was granted by the Texas Department of Health. The Houston solid waste study extended from 1979–1984 – when the *Bean v. Southwestern Waste Management, Corp.* case went to trial (Bullard, 1983, 1987).

In order to obtain the history of waste disposal facility siting in Houston, government records (city, county, and state documents) had to be manually retrieved, because the files were not computerized in 1979. On-site visits, windshield surveys, and informal interviews – done in a sort of "sociologist as detective" role – were conducted as a reliability check. This was not an easy task. I often joke about the research being done "BC" (Before Computers). It was also initiated pre-GIS (Geographic Information System).

Houston is basically flat, with many low-lying sections below sea level. In our search for landfills, we discovered a few "mountains." Anytime we found a mountain in Houston, we immediately became suspicious. They usually turned out to be old landfills. Houston also has more than 500 neighborhoods, spread over 600 square miles. Houston's black population is located in a broad belt that extends from the south-central and southeast portions of the city into northeast and north-central sections. It is also the only major American city without zoning. This no-zoning policy allowed for an erratic land-use pattern. The NIMBY (not in my backyard) practice was replaced with the "PIBBY" (place in blacks' backyard) policy (Bullard, 1983, 1987).

The all-white, all-male city government and private industry targeted landfills, incinerators, garbage dumps, and garbage transfer stations for Houston's black neighborhoods. Clearly, white men decided Houston's garbage dumps were not compatible with the city's white neighborhoods. Having white women on the city council made little difference on Houston's landfill siting in black neighborhoods. This idea changed somewhat when the first African American, Judson Robinson, Jr., was elected to the city council in 1971. Robinson had to quell a near-riot over the opening of landfill in the predominately black Trinity Gardens neighborhood (Bullard, 1987, pp. 72–73).

Five decades of discriminatory land-use practices lowered black residents' property values, accelerated physical deterioration, and increased disinvestment in Houston's black neighborhoods. Moreover, the discriminatory siting of solid waste facilities stigmatized the black neighborhoods as "dumping grounds" for a host of other unwanted facilities, including salvage yards, recycling operations, and automobile "chop shops."

Ineffective land-use regulations have created a nightmare for many of Houston's neighborhoods – especially the ones that are ill-equipped to fend off industrial encroachment. From the 1920s through the 1970s, the siting of nonresidential facilities heightened animosities between the black community and the local government. This is especially true in the case of solid waste disposal siting. In 1983, I published my first article on the Houston case detailing the proliferation of waste site in Houston's black neighborhoods (Bullard, 1983). A new environmental justice theme emerged around the idea that "since everybody produces garbage, everybody should have to bear the burden of garbage disposal." This principle made its way into the national environmental justice movement and later thwarted subsequent waste facility siting in Houston.

Public officials learned fast that solid waste management can become a volatile political issue. Generally, controversy centered around charges that disposal sites were not equitably spread in quadrants of the city. Finding suitable sites for sanitary landfills has become a critical problem, mainly because no one wants to have a waste facility as a neighbor. The burden of having a municipal landfill, incinerator, transfer station, or some other type of waste disposal facility near one's home has not been equally borne by Houston's neighborhoods.

It is clear that *Bean v. Southwestern Waste Management, Inc.* exposed the racist waste facility siting practices in Houston. The lawsuit also changed the city's facility siting and solid waste management practices after 1979 to present – forcing the city to adopt a more aggressive waste minimization and recycling plan. Since the Bean lawsuit, not a single landfill has been sited in Houston.

The national environmental justice movement was born in the 1980s in rural mostly black Warren County, North Carolina – where more than 500 blacks and whites, young and old, went to jail protesting the placement of a toxic waste dump in the mostly black and poor county. Although the demonstrations in North Carolina were not successful in halting the landfill construction, the protests brought a sharper focus to the convergence of civil rights and environmental rights and mobilized a nationally broad-based group to protest these inequities.

For me, Warren County made it clear that the Houston landfill case was not an isolated problem. This fact was later reinforced by the U.S. General Accounting Office (1983) report on hazardous waste facility siting in the South and the groundbreaking *Toxic Wastes and Race* report from the United Church of Christ Commission for Racial Justice, 1987).

Not wanting to go it alone, I recruited my friend and colleague Beverly Wright (whom I met at a Mid-South Sociological Association Meeting in

1977 in Monroe, Louisiana), who was also a young untenured sociology professor at the University of New Orleans (she was the only black female professor in a tenure-track position at that time), to work with me on environmental justice and environmental racism. I convinced her to switch from teenage pregnancy and social psychology to environment and social justice. I now had a sociologist colleague to pursue a line of research in yet untested waters. We were able to collaborate on a half dozen articles and scholarly papers in the mid-1980s.

ANATOMY OF DUMPING IN DIXIE

After the *Bean* lawsuit was resolved and appeals ended, in 1985, and after the publication of *Toxic Wastes and Race* in 1987, I expanded my analysis from Houston to examining the location of lead smelters in Dallas, a hazardous waste incinerator in Alsen, Louisiana, a toxic waste landfill in Emelle, Alabama, and a chemical plant in Institute, West Virginia. I received a grant from the National Science Foundation and Resources for the Future (RFF) to support this research – the basis for *Dumping in Dixie: Race, Class and Environmental Quality* – the first environmental justice book (Bullard, 2000).

Many residents came to understand the research I was conducting and to appreciate the noble profession of sociology as a field in which grandiose theories are developed, hypotheses formulated, and data collected that result in the verification of the obvious: Most residents of segregated black Houston neighborhoods not only knew which days the garbage was collected but also knew the addresses of the existing and abandoned landfills and incinerators. Many of these same residents had spent much of their lives escaping from waste sites, only to find waste facility disputes following them to their new neighborhoods.

In *Dumping in Dixie*, I explored the thesis that African American communities in the South – the nation's Third World – because of their economic and political vulnerabilities, were routinely targeted for the siting of noxious facilities, LULUs, and environmental hazards. People in these communities, in turn, were also likely to suffer greater environmental and health risks than in the general population.

The issues addressed in *Dumping in Dixie* center on equity, fairness, and the struggle for social justice by African American communities. The struggles against environmental injustice are not unlike the civil rights battles waged to dismantle the legacy of Jim Crow in Selma, Montgomery,

Birmingham, and some of the "Up South" communities in New York, Boston, Philadelphia, Detroit, Chicago, and Los Angeles. The analysis chronicles the environmental justice movement in an effort to develop common strategies that are supportive of building sustainable communities by and for African Americans and other people of color.

In the South, African Americans make up the region's largest racial minority group. This analysis could have easily focused on Latino Americans in the Southwest or Native Americans in the West. People of color in all regions of the country bear a disproportionate share of the nation's environmental problems. Racism knows no geographic bounds.

Limited housing and residential options, combined with discriminatory facility practices, have contributed to the imposition of all types of toxins on African American communities through the siting of garbage dumps, hazardous-waste landfills, incinerators, smelter operations, paper mills, chemical plants, and a host of other polluting industries. These industries have generally followed the path of least resistance, which has been to locate in economically poor and politically powerless African American communities.

I found that poor African American communities were not the only victims of facility siting disparities and environmental discrimination, however. Middle-income African American communities were also confronted with many of the same land-use disputes and environmental threats as their lower-income counterparts. Increased income has enabled few African Americans to escape the threat of unwanted land uses and potentially harmful environmental pollutants. In the real world, racial segregation is the dominant residential pattern and racial discrimination is the leading cause of segregated housing in America.

Since affluent, middle-income, and poor African Americans live within close proximity to one another, the question of environmental justice can hardly be reduced to a poverty issue. The black middle-class community members in Houston's Northwood Manor neighborhood quickly discovered that their struggle was not unlike that of their working-class and poor counterparts who had learned to live with that city's garbage dumps and incinerators. For those making environmental and industrial decisions, African American communities – regardless of their class status – were considered to be throw-away communities, compatible with garbage dumps, transfer stations, incinerators, and other waste disposal facilities.

We were beginning to see a growing number of African American and people of color grassroots activists challenge public policies and industrial practices that threaten the residential integrity of their neighborhoods.

Activists began to demand environmental justice and equal protection. The demands were reminiscent of those voiced during the civil rights era – they were for an end to discrimination in housing, education, employment, and the political arena. Many exhibited a growing militancy against industrial polluters and government regulatory agencies that provided theses companies with permits and licenses to pollute.

The national environmental justice movement had not yet been discovered by the white media. I knew it was just a matter of time. The movement was still emerging. Thus, getting *Dumping in Dixie* published was not an easy task. I was rejected by dozens of publishers who basically said "there is no connection between race and the environment" before Westview Press agreed to publish the book in 1990. That same year, the National Wildlife Federation (NWF) gave me the Conservation Achievement Award in Science for the book and invited me to spend the summer in Washington, DC as a visiting scholar.

NWF even put me and my family up in a swank apartment near their headquarters in Washington, DC. That summer my wife and kids had a ball doing the Washington, DC, "tourist thing," while I gave a couple of lectures and schooled NWF staffers on environmental justice and the importance of diversifying its staff, board, and agenda.

On the other hand, my colleagues in the American Sociological Association were slow to recognize environmental justice as a movement. The Environment and Technology Section did give me an award in 1998 for *Dumping in Dixie*, a full eight years after it was first published. My position was better late than never. I reluctantly accepted the award – not that I needed the ASA or the NWF to validate my EJ work or the existence of the EJ Movement.

EJ IN THE 1990s

The 1990s saw the terms "environmental justice," "environmental racism," and "environmental equity" becoming household words. Out of the small and seemingly isolated environmental struggles emerged a potent grassroots movement. Environmental justice became a unifying theme across race, class, gender, age, and geographic lines.

In the 1990s, many Americans, ranging from constitutional scholars to lay grassroots activists, came to recognize that environmental discrimination is unfair, unethical, and immoral. The practice is also illegal. I carried all of my research with the assumption that all Americans have a basic right to live, work, play, go to school, and worship in a clean and healthy environment.

This framework became the working definition of the environment for many environmental justice activists and analysts alike. I made a deliberate effort to write readable books that might reach a general audience while at the same time covering uncharted areas of interest to environmentalists, civil rights advocates, community activists, political leaders, and policymakers.

The decade of the 1990s was a different era from the late1970s. Some progress was made in mainstreaming environmental protection as a civil rights and social justice issue. In the early 1990s, I began to get calls from all kinds of groups wanting me to speak at forums, symposia, and conferences. Many of these groups covered travel expenses, and some even paid me to speak. I was getting so many calls in 1991 that my colleagues Beverly Wright jokingly suggested I get a booking agent. I took her advice and hired Jodi Solomon Speakers Bureau in Boston to handle my speaking engagements. Jodi sure has made my life easier.

The 1990s also saw groups like the NAACP Legal Defense and Education Fund, Earthjustice Legal Defense Fund, Lawyers Committee for Civil Rights Under the Law, Center for Constitutional Rights, National Lawyers Guild's Sugar Law Center, American Civil Liberties Union, and Legal Aid Society team up on environmental justice and health issues that differentially affect poor people and people of color. Environmental racism and environmental justice panels became "hot" topics at conferences sponsored by law schools, bar associations, public health groups, scientific societies, and social science meetings.

Environmental justice leaders have also had a profound impact on public policy, industry practices, national conferences, private foundation funding, and academic research. Environmental justice courses and curricula can be found at nearly every university in the country. The 1990s saw a growing number of academics building careers, getting tenure, promotion, and merit raises in environmental justice.

Environmental justice trickled up to the federal government and the White House. Environmental justice activists and academicians were key actors who convinced the U.S. Environmental Protection Agency (under the Bush Administrations) to create an Office of Environmental Equity. In 1990, the Reverend Benjamin Chavis (who at the time was Executive Director of the United Church of Christ Commission for Racial Justice) and I were selected to work on President Bill Clinton's Transition Team in the Natural Resources Cluster (the EPA, and the Departments of Energy, the Interior, and Agriculture).

The 1991 First National People of Color Environmental Leadership Summit was probably the most important single event in the movement's history. The Summit, coordinated by the New York-based United Church of Christ Commission for Racial Justice, broadened the environmental justice movement beyond its early anti-toxics focus to include issues of public health, worker safety, land use, transportation, housing, resource allocation, and community empowerment. As a member of the National Planning Committee, I was flying to New York twice a month for meetings.

The meeting also demonstrated that it is possible to build a multi-racial grassroots movement around environmental and economic justice. Held in Washington, DC, the four-day summit was planned for 500 participants – more than 1,000 attended from around the world. Delegates came from all fifty states, including Alaska and Hawaii, as well as Puerto Rico, Chile, Mexico, and as far away as the Marshall Islands. People attended the Summit to share their action strategies, redefine the environmental movement, and develop common plans for addressing environmental problems affecting people of color in the United States and around the world.

On September 27, 1991, Summit delegates adopted 17 "Principles of Environmental Justice." These principles were developed as a guide for organizing, networking, and relating to government and nongovernmental organizations (NGOs). By June 1992, Spanish and Portuguese translations of the Principles were being used and circulated by NGOs and environmental justice groups at the Earth Summit in Rio de Janeiro. I was surprised to see a young Brazilian woman reciting our principles in the Global Forum.

Many grassroots activists are convinced that waiting for the government to act has endangered the health and welfare of their communities. Unlike the federal EPA, communities of color did not first discover environmental inequities in 1990. The federal EPA only took action on environmental justice concerns in the 1990s after extensive prodding from grassroots environmental justice activists, educators, and academics.

As spelled out in greater detail in Dorceta Taylor's chapter in this volume, even though environmental justice was on its way to becoming mainstream, there was still a dominant myth that poor people and people of color were not concerned about the environment because many of us were not card-carrying members of the Sierra Club, the National Audubon Society, or the other white environmental groups. Since none of our groups were listed in mainstream environment and conservation directories, including NWF's widely used *Conservation Directory*, it was assumed that we must not have any groups of our own that work on these issues.

Because of this gross oversight, I produced the *People of Color Environmental Groups Directory in the United States* in 1992. The preliminary listings were used as the invitation mailing list for the First National People of Color Environmental Leadership Summit, held in Washington, DC in 1991. The 2nd edition of the *Directory* was published in 1994 and included groups in the United States, Canada, and Mexico. The 3rd edition of the *Directory* was published in 2000. The number of groups grew from 300 in 1992 to over 1,000 listings in 2000.[5]

During the 1991 Summit, several of my colleagues and I who were at majority-white universities worked on a strategy to establish environmental justice centers at historically black colleges and universities (HBCUs) in the South. In 1992, Beverly Wright left her position at Wake Forest University to start the Deep South Center for Environmental Justice at Xavier University in New Orleans.

In 1994, I left the University of California at Riverside to establish the Environmental Justice Resource Center at Clark Atlanta University. Two other environmental justice centers were established that year: the Thurgood Marshall Environmental Justice Legal Clinic (Texas Southern University-Houston, TX), and the Environmental Justice and Equity Institute (Florida A&M University-Tallahassee, FL).

It is no coincidence that in 1994, Bunyan Bryant and Paul Mohai were able to get environmental justice courses approved at the University of Michigan School of Natural Resources and Environment, setting the stage for that university to become the nation's first and only academic program to offer bachelor's, master's, and doctoral degrees in environmental justice.

As members of the Clinton Transition Team, Reverend Ben Chavis and I were able to convey the priorities that were given to us by EJ leaders from across the country. EJ leaders asked the Clinton administration to fully fund and staff EPA's Office of Environmental Equity, issue an Executive Order on EJ, establish an EJ advisory council under the Federal Advisory Committee Act (FACA), and appoint people of color to half of the EPA's 10 regions and the Natural Resources Cluster (EPA, Department of Energy, Department of Interior, Department of Agriculture).

The Clinton Administration made good on several other EJ priorities, including the appointment of African Americans to head the Department of Energy (Hazel O'Leary) and Department of Agriculture (Mike Espy); EPA's Office of Environmental Equity received increased funding and staff, and the name was changed to Office of Environmental Justice. In addition, the National Environmental Justice Advisory Council (NEJAC) was established to advise EPA.

In response to growing public concern and mounting scientific evidence, the National Institute of Environmental Health Sciences (NIEHS) convened a "Health and Research Needs to Ensure Environmental Justice Symposium" in Crystal City, Virginia, and as noted in the introduction to this chapter, on the second day of that symposium (February 11, 1994y), President Clinton issued Executive Order 12898, "Federal Actions to Address Environmental Justice in Minority Populations and Low-Income Populations." We were attending the symposium when we got the call from the White House. A half dozen of us were invited to the oval office on that cold snowy day in February 1994 when President Bill Clinton signed the EJ Executive Order. Clinton gave us all pens and later a photograph of the historic signing.

The Executive Order attempts to address environmental injustice within existing federal laws and regulations. It reinforces the Civil Rights Act of 1964, Title VI, which prohibits discriminatory practices in programs receiving federal funds. The Order also focuses the spotlight back on the National Environmental Policy Act (NEPA), a thirty-five-year-old law that set policy goals for the protection, maintenance, and enhancement of the environment. NEPA's goal is to ensure for all Americans a safe, healthful, productive, and aesthetically and culturally pleasing environment. NEPA requires federal agencies to prepare a detailed statement on the environmental effects of proposed federal actions that significantly affect the quality of "the human environment."

The Executive Order calls for improved methodologies for assessing and mitigating impacts, health effect from multiple and cumulative exposure, collection of data on low-income and minority populations who may be disproportionately at risk, and impacts on subsistence fishers and wildlife consumers. It also encourages participation of the impacted populations in the various phases of assessing impacts – including scoping, data gathering, identification of alternatives, analysis, mitigation, and monitoring.

EJ IN THE NEW MILLENNIUM

In August 2000, I was part of a delegation of environment justice and human rights leaders attending the United Nations Conference Against Racism (WCAR) in Durban, South Africa, that joined the NGO Environmental Justice Caucus to help establish environmental racism as a human rights violation. The Caucus was successful in advancing language in the NGO Declaration and Program of Action, urging governments to adopt policies that will lead to the "eradication of environmental racism" and quicken the "quest for sustainable development." The NGO language also encourages governments to implement policies that will protect the environment and establishes a fundamental right to clean air, land, water, food, and safe and decent housing.

Two years later, in 2002, I joined an eight-person National Black Environmental Justice Network (NBEJN) delegation that participated in the World Summit on Sustainable Development (WSSD) Preparatory Committee (Prepcom) IV meeting held in Bali, Indonesia. The meeting ran from May 28 through June 7, 2002, and was attended by government and nongovernmental leaders from around the world. As part of a delegation sponsored by the Ford Foundation, NBEJN members also linked up with other environmental justice and social justice advocates from around the world.

Because of the frustration and disappointment with the Precom IV process and expected outcomes, the Sustainable Development Issues Network (SDIN) held an NGO press conference of a dozen caucuses. The caucuses also released a collective statement, "What on Earth is Missing?" They concluded that Bali has failed glaringly in addressing issues related to rebalancing the economic context of environmental degradation and fundamental inequities of the international economic system.

More than 300 environmental justice leaders from around the world gathered at the Shaft 17 Education Center in Johannesburg to participate in the Environmental Justice Forum. The four-day forum, sponsored by the South African-based Environmental Justice Networking Forum (EJNF), served as a pre-summit kick-off to the opening of the WSSD and the Global People's Forum – a meeting of nongovernmental organizations that run parallel to the official government meeting. The meetings ran from August 26 through September 4, 2002 in Johannesburg.

The EJNF forum examined the costs and consequences of environmental racism and nonsustainable development practices and their contribution to poverty, pollution, and ill health. Many of the South Africans blacks are the "poorest of the poor," that one-fifth of the world's population living on less than $1 a day and unable to secure adequate food, water, clothing, shelter, and health care. Thabo Madihlaba, national coordinator of the EJNF, told me his group is engaged in an everyday struggle to recover from the legacy of apartheid and environmental racism. Lack of affordable energy, sanitation, and access to clean water pose severe health threats to millions of South Africans who remain crowded in substandard housing.

Longtime South African human rights activist Dennis Brutus criticized the slow pace of the government getting basic services to its people. "The legacy of apartheid is still with us and accounts for poor health delivery, inadequate water and sanitation infrastructure, poor housing, and lack of basic services such as electricity," Brutus told the EJNF forum. Almost 15 million black South Africans out of a total population of 43.5 million are without electricity and live on less than $2 per day.

Inadequate sanitation accounts for over 43,000 South African children deaths each year. Getting clean and affordable water delivered to the people is a major challenge. Over 10 million South Africans had their water cut off. Still, these problems are not limited to South Africa: More than 1.4 billion people around the world lack access to safe water. Dirty water is one of the world's "deadliest" pollutants.

Presenter after presenter challenged government and transnational corporations to halt their destructive and nonsustainable practices that harm the poor, people of color, and indigenous peoples. My EJ colleagues and I came to Johannesburg and the WSSD to share our experiences and work with our friends in South Africa and with other groups around the world to craft common strategies to build a movement for achieving safe, healthy, just, and sustainable communities.

South Africa is grappling with the legacy of apartheid that herded approximately 87 percent of the black population into 13 percent of the country's land. Land redistribution is a core environmental, economic, and political issue. Just days before the official opening of the WSSD meetings, the South African government arrested 72 members of the Landless Peoples Movement (LPM) for staging a protest in protest at the John Vorster Square. LPM was protesting the eviction of people who were living on government land due to apartheid settlement policies.

EJNF and NBEJN joined with the LPM in calling on the South African government to complete the land redistribution scheme in an appropriate and timely process. Landlessness is tantamount to structural apartheid and a manifestation of environmental racism, which is reprehensible in a free and democratic society. The World Summit on Sustainable Development could have been a showcase for real sustainable development rather than being hijacked by the transnational corporations.

Shortly after the WSSD, I served on the Executive Committee for the planning of the Second National People of Color Environmental Leadership Summit, which was held in Washington, DC on October 23–26, 2002. We planned the meeting for 500 delegates. Over 1,400 participants attended the four-day event. In order to have substantive resource materials going into and coming out of the EJ Summit II, my center coordinated a national "Call for Policy Papers" on a wide range of topics, including environmental justice, transportation equity, children health, renewable energy, suburban sprawl, and regional equity, livable communities, and related topics resulting in over two dozen resource/policy papers.

We collected 26 policy papers during the Summer 2002 and distributed at the EJ Summit II. The policy papers were written by the nation's leading academics, policy analysts, legal scholars, health experts, practitioners, and activists. They were designed for use in the EJ Summit II general sessions, workshops, hands-on training, and meeting deliberations.

GOVERNMENT RESPONSE TO WEATHER-RELATED DISASTERS

Each year, communities along the Atlantic and Gulf Coast states are hit with tropical storms and hurricanes forcing millions to flee to higher ground. Institutionalized racism in housing and land use planning also provides privilege for whites in securing the higher ground and environmentally safer neighborhoods. Where whites choose to live, work, play, go to school, and worship is not accidental (Lareau, 2000). Many of their choices are shaped by race.

Historically, the Atlantic hurricane season produces 10 storms, of which about six become hurricanes and two to three become major hurricanes. However, the 2005 hurricane season produced a record 27 named storms – topping the previous record of 21 storms, set in 1933 – and 13 hurricanes, besting the old record of 12 hurricanes, set in 1969 (Tanneeru, 2005), as the most hurricanes in one season since record keeping began in 1851 (Cuevas, 2005).

Government has a long history of discriminating against black victims of hurricanes, floods, and droughts. Clearly, race matters in terms of swiftness of response, allocation of post-disaster assistance, and reconstruction assistance. Emergency response often reflects the pre-existing social and political stratification structure, with black communities receiving less priority than white communities. Race and class dynamics play out in disaster survivors' ability to rebuild, replace infrastructure, obtain loans, and locate temporary and permanent housing (Bolin & Bolton, 1986; Dyson, 2006; Pastor et al., 2006).

AVERTING A SECOND DISASTER – KATRINA AND THE "TWENTY-POINT" PLAN

On August 29, 2005, Hurricane Katrina made landfall near New Orleans, leaving death and destruction across the Louisiana, Mississippi, and Alabama Gulf Coast. Katrina is likely the most destructive hurricane in the U.S. history, costing over $70 billion in insured damage (Brinkley, 2006; Van Heerden, 2006; Hartman & Squires, 2006). Katrina was also one of the deadliest storms in decades, with a death toll of at least 1,325. The death toll of this storm is surpassed only by the 1928 hurricane in Florida, where estimates vary from 2,500 to 3,000, and the 8,000 deaths recorded in the 1900 Galveston hurricane (Kleinberg, 2003).

On December 23, 2005, two days before Christmas and three months after Hurricane Katrina made landfall in Louisiana and the levee breech drowned New Orleans, I wrote a short essay, "Katrina and the Second Disaster: A Twenty-Point Plan to Destroy New Orleans" ("Twenty-Point Plan") that examined some key policies and practices being put in place to address the worse disaster in the U.S. history. The "Twenty-Point Plan" was based on trends and observations Beverly Wright, a native New Orleanian who evacuated to Atlanta after Katrina, and I, discussed daily at my center. I was on sabbatical and had time to devote exclusively to Katrina issues.

Shortly after the storm, I helped Wright set up a temporary office for her center at Clark Atlanta University – because the EJ center she founded at Xavier University (her center was scheduled to move to Dillard University on September 1, 2005) in New Orleans went under with the flood. I agreed to synthesize these disturbing trends, policies, proposals, and events points into a list. Twenty points sounded like a nice round number for this list.

As reconstruction and rebuilding moved forward in New Orleans, it was clear that the lethargic and inept emergency response after Katrina was a disaster that overshadowed the deadly storm itself. A post-storm "second disaster" is now driven by racism, classism, elitism, paternalism, and old-fashion greed (Bullard, 2005c). The elements in the "Twenty-Point plan" are as listed below.

1. Selectively hand out FEMA grants.

    The FEMA is being consistent in the slow response in getting aid to Katrina survivors. FEMA's grant assistance program favors middle-income households. A second disaster is created by making it difficult for low-income and black Katrina survivors to access government assistance and directing the bulk of the grant assistance to middle-income white storm victims. The Lawyers Committee for Civil Rights and several other legal groups have sued FEMA over its response and handling of aid to storm victims. FEMA has referred more than 2 million people, many of them with low incomes, to the Small Business Administration (SBA) to get the loans.

2. Systematically deny the poor and blacks SBA loans.

The man-made second disaster can be exacerbated by screening out the poor and denying black households disaster loans. The *New York Times* editorial summed up this problem: "The Poor Need Not Apply." The SBA has processed only a third of the 276,000 home loan applications it has received. However, the SBA has rejected 82 percent of the applications it received, a higher percentage than in most previous disasters. Well-off neighborhoods like Lakeview have received 47 percent of the loan approvals, while poverty-stricken neighborhoods have gotten 7 percent. Middle-class black neighborhoods in the eastern part of the city have lower loan rates.

3. Award insurance claims using the "wind or water" trap.

Because of the enormity of the damage in the wake of Katrina, insurance companies will categorize a lot of legitimate wind claims as flood- or water-related. The "wind or water" problem will hit black storm victims hardest because they are likely to have their insurance with small companies – since the major firms "redlined" many black neighborhoods. Most rebuilding funds after disasters come from private insurance – not the government.

4. Redline black insurance policyholders.

Numerous studies [where are the references?] show that African Americans are more likely than whites to receive insufficient insurance settlement amounts. Insurance firms target black policyholders for low and inadequate insurance settlements based on majority black zip codes to subsidize fair settlements made to white policyholders. If black homeowners and business owners expect to recover from Katrina, then they must receive full and just insurance settlements. FEMA and the SBA cannot be counted on to rebuild black communities.

5. Use "green building" and flood-proofing codes to restrict redevelopment.

Requiring rebuilding plans to conform to "green building" materials and new flood-proofing codes can price many low- and moderate-income homeowners and small business owners out of the market. This will hit black homeowners and black business owners especially hard since they generally have lower incomes and lower wealth.

6.   Apply discriminatory environmental clean-up standards.

     Failure to apply uniform clean-up standards can kill off black
neighborhoods. Environmental racism is operationalized in the full-scale
cleanup of white neighborhoods to residential standards, while allowing
no cleanup or partial cleanup – industrial standards – of black residential
neighborhoods. Failure to clean up black residential areas can act
as a disincentive for redevelopment. It could also make people sick.
An environmental injustice occurs when the government uses the
argument that black neighborhoods were already highly polluted with
background contamination, or "hot spots," exceeding EPA safe levels
pre-Katrina and thus need not be cleaned to more rigorous residential
standards.

7.   Sacrifice "low-lying" black neighborhoods in the name of saving the
     wetlands and environmental restoration.

     Allow black neighborhoods like the Lower Ninth Ward and New
Orleans East to be "yielded back to the swamp" while allowing similar
low-lying white areas to be rebuilt and redeveloped. This is a form
of "ethnic cleansing" that was not possible before Katrina. Instead of
emphasizing equitable rebuilding, uniform clean-up standards, equal
protection, and environmental justice for African American communities, public officials should send mixed signals for rebuilding vulnerable
"low-lying" black neighborhoods.

8.   Promote a smaller, more upscale, and "whiter" New Orleans.

     Concentrating on getting less-damaged neighborhoods up and running
could translate into a smaller, more upscale, and whiter New Orleans and
a dramatically down-sized black community. Clearly, shrinking New
Orleans neighborhoods disproportionately shrinks black votes, black
political power, and black wealth.

9.   Revise land use and zoning ordinances to exclude.

     Katrina can be used to change land use and zoning codes to "zone
against" undesirable land uses that were not politically possible before
the storm. Also, "expulsive" zoning can be used to push out certain land
uses and certain people.

10. Phased rebuilding and restoration scheme that concentrates on the
     "high ground."

     New Orleans officials are being advised to concentrate rebuilding on
the areas that remained high and dry after Katrina. These areas are
disproportionately white and affluent. This scenario builds on pre-
existing inequities and "white privilege" and ensures future inequities and
"white privilege." By the time rebuilding gets around to black "low-

lying" areas, there are not likely to be any rebuilding funds left. This is the "oops, we are out of funds" scenario.

11. Apply eminent domain as a black land grab.

Give Katrina evacuees one year to return before the city is allowed to legally "take" their property through eminent domain. Clearly, it will take much longer than a year for most New Orleanians to return home. This proposal could turn into a giant land grab of black property and loss of black wealth they have invested in their homes and businesses.

12. No financial assistance for evacuees to return.

Thousands of Katrina evacuees were shipped to more than three dozen states with no provisions for return – equivalent to a "one-way" ticket. Many Katrina evacuees are running short of funds. No money translates into no return to their homes and neighborhoods. Promoting the "right to return" without committing adequate resources to assist hurricane evacuees to return is mere political rhetoric.

13. Keep evacuees away from New Orleans jobs.

The nation's unemployment rate was 5 percent in November 2005. The November 2005 jobless rate for Katrina returnees was 12.5 percent, while 27.8 percent of evacuees living elsewhere were unemployed. However, the black jobless rate was 47 percent in November, compared with 13 percent for whites who have not gone back (Uchitelle, 2005).

Katrina evacuees who have made it back to their home region have much lower levels of joblessness. This is especially important for African Americans, whose joblessness rate fell over 30 percentage points for returnees. The problem is that the vast majority of black Katrina evacuees have not returned to their home region. Only 21 percent of black evacuees have returned, compared with 48 percent of whites.

14. Fail to enforce fair housing laws.

Allowing housing discrimination against blacks to run rampant is an artificial roadblock to repopulation of black New Orleans. Katrina created a housing shortage and opened a floodgate of discrimination against black homeowners and renters. In December 2005, the National Fair Housing Alliance (2005) found high rates of housing discrimination against African-Americans displaced by Hurricane Katrina. In 66 percent of the tests conducted by the NFHA, 43 of 65 instances, whites were favored over African-Americans.

15. No commitment to rebuild and replace low-income public housing.

Shortly after Katrina struck, even the secretary of the U.S. Department of Housing and Urban Development (HUD) spoke of not rebuilding all of the public housing lost during the storm. The HUD secretary's statement sent a powerful signal to New Orleans' poor that public housing may not be around for them to return to.

16. Downplay the black cultural heritage of New Orleans.

Promote rebuilding and the vision of a "new" New Orleans as if the rich black culture did not matter or act as if it can be replaced or replicated in a "theme park"-type redevelopment scenario. Developers should capture and market the "black essence" of New Orleans without including black people.

17. Treatment of mixed-income housing as superior to all-black neighborhoods.

First, there is nothing inherently inferior about an "all-black" neighborhood – or an all-black anything for that matter. Black New Orleanians who chose to live in neighborhoods that happened to be all-black – just as whites have always had the right to move in or move out of these neighborhoods – should not be forced to have their neighborhoods rebuilt as "integrated" or "multicultural" neighborhoods. Also, "mixed-income" housing, to many blacks, conjures up the idea of 10 percent of the fair market housing units set aside for them. Many blacks are battle-weary of competing for that 10 percent. New Orleans was 68 percent black before Katrina – and most black folks were comfortable with that.

18. Allow "oversight" (overseer) board to manage Katrina funds that flow to New Orleans.

Take away "home rule," since the billions of Katrina redevelopment dollars that will flow to New Orleans is too much money for a majority black city council and a black mayor to oversee or manage. More important, the oversight board will need to represent "big-money" interests- real estate, developers, banking, insurance, hotels, law firms, tourist industry etc. – well beyond the purview of a democratically elected city government to ensure that the vision of the "new" New Orleans, "smaller and more upscale," gets implemented.

19. Delay rebuilding and construction of New Orleans schools.

The longer the New Orleans schools stay closed, the longer the families with children will stay away. Schools are a major predictor of racial polarization. Before Katrina, over 125,000 New Orleans children were attending schools in the city. Blacks made up 93 percent of the students in New Orleans schools. Evacuated children are enrolled in school districts from Arizona to Pennsylvania. Three months after the storm, only one of the New Orleans' 116 schools was open.

20. Hold elections without appropriate Voting Rights Act safeguards.

Almost 300,000 registered voters left New Orleans after Katrina. The powerful storm damaged or destroyed 300 of the 442 polling places. Holding city elections pose major challenges regarding registration, absentee ballots, city workers, polling places, and identification for displaced New Orleanians. Identification is required at the polls, and returning residents may not have access to traditional identification papers – birth certificates, drivers' licenses, etc. – destroyed by the hurricane. More than three months after Katrina struck, 80 percent of New Orleans voters had not made their way back to the city, including most African Americans, who comprised a two-third majority of the population before the storm. Most of the estimated 60,000–100,000 New Orleans residents who made it back by the end of 2005 were white and middle-class, changing the racial and political complexion of the city. Holding elections while the vast majority of New Orleans voters are displaced outside their home district and even their home state is unprecedented in the history of the United States, but it also raises racial justice and human rights questions.

Eighteen months after the storm, it is clear that major elements of the "Twenty-Point Plan" have gradually been implemented – whether by design or by default. Katrina floodwaters may have swept New Orleanians from their city, but the politics of race is keeping most African American evacuees from returning home. This brutal fact is made clear by a steady stream of empirical studies – ranging from repopulation projections of a smaller and "whiter" New Orleans footprint, racial barriers to government loans and grants, insurance "tug of war" centered around "wind vs. water," racial redlining, inadequate clean-up standards, redevelopment plans that permanently displaces poor black families from public housing developments (Pastor et al., 2006; Wright & Bullard, 2007a, 2007b).

HUD announced it would invest $154 million in rebuilding public housing in New Orleans and assist the city to bring displaced residents home. But critics fear that government officials and business leaders are quietly planning to demolish the old projects and privatize public housing. Before the storm, New Orleans had close to 8,000 public housing units. A year after Katrina, at least 80 percent of public housing in New Orleans remained closed. Six of ten of the largest public housing developments in the city were boarded up, with the other four in various states of repair.

Over 49,000 people lived in public housing before Katrina, 20,000 in older, large-scale developments such as St. Bernard, and 29,000 in Section 8 rental housing, which was also devastated by the storm. Although the city faces a severe housing crunch, in June 2006, federal housing officials announced that more than 5,000 public housing apartments for the poor would be razed and replaced by developments for residents from a wider range of incomes (Saulny, 2006). This move heightened the anxiety of many low-income black Katrina survivors who fear they will be pushed out in favor of higher-income families.

The demolition of four sprawling public housing projects – the St. Bernard, C.J. Peete, B.W. Cooper, and Lafitte housing developments – represents more than half of all of the conventional public housing in the city, where only 1,097 units were occupied 10 months after the storm. An unresolved lawsuit on behalf of residents charges that the demolition plan is racially discriminatory. HUD raised by 35 percent the value of disaster-vouchers for displaced residents because the city's housing shortage has caused rents to skyrocket.

Many, if not most, Katrina victims may not have resources to hire lawyers to fight the insurance companies. Although neighborhoods of the poor and people of color suffered the brunt of Katrina and the subsequent flooding, residents living in white neighborhoods have been three times as likely as homeowners in black neighborhoods to seek state help in resolving insurance disputes.

A January 2007 Associated Press (AP) analysis of Louisiana's insurance complaints settled in the first year after Katrina exposes the cold, hard truth about Katrina's winds and waters argument (Callimachi & Bass, 2006). The AP report concludes that people of color, who often need the most help after a major disaster, are the ones who are most disconnected from the government institutions that can provide it. This problem is compounded by their distrust of those in power (Callimachi & Bass, 2006).

Many African American Katrina survivors who lost their homes are simply giving up or are taking what their insurer gave them. Few blacks even knew they could appeal to the state. However, even having that information does not mean that African Americans trust the state of Louisiana or have confidence in the state to be fair – given the history of differential treatment of blacks at the hands of the state government.

In October 2006, more than 700,000 insurance claims were filed for damage resulting from Katrina in Gulf Coast states, and only $14.9 billion out of $25.3 billion insured losses had been paid. In Louisiana, more than 8,000 residents had filed Katrina-related complaints with the state insurance office. Using open records law, the AP obtained the files of more than 3,000 complaints that have already been settled and analyzed the outcomes by the demographics of the victims' current zip code neighborhood.

The AP found that nearly 75 percent of the settled cases were filed by residents currently living in predominantly white neighborhoods. Only 25 percent were filed by households in majority-black zip codes. The analysis also suggests income was a factor – with the average resident who sought state help living in a neighborhood with a median household income of $39,709, compared with the statewide median of $32,566 in the 2000 Census. These findings surprise few on the front lines of a disaster that has reawakened issues of racial equality.

In an attempt to head off a flood of insurance disputes, Mississippi Attorney General Jim Hood filed suit to block insurance companies from denying flood claims when those floods are caused by wind. He claims that the insurance exclusion of water damage violates Mississippi's Consumer Protection Act and "deprives consumers of any real coverage choices" (Lee, 2005). The Mississippi lawsuit also accuses some insurance companies of forcing storm victims into signing documents that stipulate their losses are flood-related, not wind-related, before they can receive payment or emergency expenses; the lawsuit would ban such practices, which are tantamount to economic blackmail.

In January 2007, a jury awarded $2.5 million in punitive damages to Norman and Genevieve Broussard, who sued State Farm Fire & Casualty Co. for denying their claim after Hurricane Katrina – a decision that could benefit hundreds of other homeowners challenging insurers for refusing to cover billion of dollars in storm damage. The Broussards sued State Farm after it refused to pay when their Biloxi home was destroyed by Katrina's winds and subsequent storm surge from the Gulf of Mexico (Associated 2007).

The U.S. District Judge L.T. Senter Jr. found State Farm's position so incredible that he took the case out of the hands of the jury and ruled against the company before handing the case back to the jury for a determination on punitive damages. Judge Senter said that there was no legal basis for the insurer refusing to pay the homeowners' claim. State Farm, Mississippi's largest home insurer, is now considering paying hundreds of millions of dollars to settle more than 600 lawsuits and resolve thousands of other disputed claims (Associated Press, 2007).

THE "MOTHER OF ALL TOXIC CLEANUPS": LET THEM EAT RISKS

Hurricane Katrina has been described as the worst environmental disaster in the U.S. history. The powerful storm left behind an estimated 22 million tons of debris, more than 15 times the debris hauled away from the 9/11 attack (Griggs, 2005). Half of this debris, 12 million tons, is in Orleans Parish. In addition to wood debris, EPA and LDEQ officials estimate that 140,000–160,000 homes in Louisiana may need to be demolished and disposed (Louisiana Department of Environmental Quality, 2005).

A September 2005 *Business Week* commentary described the handling of the untold tons of "lethal goop" as the "mother of all toxic cleanups." However, the billion-dollar question facing New Orleans is which neighborhoods will get cleaned up and which ones will be left contaminated (Business Week, 2005). A year after Katrina, more than 99 million cubic yards of debris have been removed in Alabama, Louisiana, and Mississippi at the cost of $3.7 billion (Federal Emergency Management Agency, 2006). Still, nearly a third of the hurricane trash in New Orleans had not been picked up a full year after the storm.

Before Katrina, the City of New Orleans was struggling with a wide range of environmental justice issues and concerns. Its location along the Mississippi River Chemical Corridor increased its vulnerability to environmental threats (Roberts & Toffolon-Weiss, 2001). The city has an extremely high childhood environmental lead poisoning problem. There were ongoing air quality impacts and resulting high asthma and respiratory disease rates and frequent visits to emergency rooms for treatment by both children and adults (Wright, 2005). Environmental health problems and issues related to environmental exposure were hot-button issues in New Orleans long before Katrina's floodwaters emptied out the city.

What gets cleaned up and where the waste is disposed are longstanding equity and environmental justice issues (Bullard, 1994). Dozens of toxic "time bombs" along Louisiana's Mississippi River petrochemical corridor, the 85-mile stretch from Baton Rouge to New Orleans, made the region a major environmental justice battleground. The corridor is commonly referred to as "Cancer Alley" (Wright, 2005). For decades, black communities all along the petrochemical corridor have been fighting against environmental racism and demanding relocation from polluting facilities.

Two mostly black New Orleans subdivisions, Gordon Plaza and Press Park, have special significance to environmental justice and emergency response (Lyttle, 2004; Wright, 2005). Both subdivisions were built on a portion of land that was used as a municipal landfill for more than 50 years. The Agriculture Street Landfill, covering approximately 190 acres, was used as a city dump as early as 1910. Municipal records indicate that after 1950, the landfill was mostly used to discard large solid objects, including trees and lumber, and it was a major source for dumping debris from the very destructive 1965 Hurricane Betsy. It is important to note that the landfill was classified as a solid waste site and not a hazardous waste site.

In 1969, the federal government created a home ownership program to encourage lower-income families to purchase their first home. Press Park was the first subsidized housing project on this program in New Orleans. The federal program allowed tenants to apply 30 percent of their monthly rental payments toward the purchase of a family home. In 1987, 17 years later, the first sale was completed. In 1977, construction began on a second subdivision, Gordon Plaza. This development was planned, controlled, and constructed by the U.S. Department of Housing and Urban Development (HUD) and the Housing Authority of New Orleans (HANO). Gordon Plaza consists of approximately 67 single-family homes.

In 1983, the Orleans Parish School Board purchased a portion of the Agriculture Street Landfill site for a school. The fact that this site had previously been used as a municipal dump prompted concerns about the suitability of the site for a school. The board contracted engineering firms to survey the site and assess it for contamination of hazardous materials. Heavy metals and organics were detected at the site.

Despite the warnings, Moton Elementary School, an $8 million "state of the art" public school, opened with 421 students in 1989. In May 1986, EPA performed a site inspection (SI) in the Agriculture Street Landfill community. Although lead, zinc, mercury, cadmium, and arsenic were found at the site, based on the Hazard Ranking System (HRS) model used at that time, the score of 3 was not high enough to place them on the National Priorities List (NPL).

On December 14, 1990, EPA published a revised HRS model in response to the Superfund Amendment and Reauthorization Act (SARA) of 1986. Upon the request of community leaders, in September 1993, an expanded site inspection (ESI) was conducted. On December 16, 1994, the Agriculture Street Landfill community was placed on the NPL with a new score of 50.

The Agriculture Street Landfill community is home to approximately 900 African American residents. The average family income is $25,000, and the educational level is high school graduate and above. The community pushed for a buy-out of their property and to be relocated. However, this was not the resolution of choice by EPA. A clean-up was ordered at a cost of $20 million, even though the community buy-out would have cost only $14 million. The actual clean up began in 1998 and was completed in 2001 (Lyttle, 2004).

Government officials assured the Agricultural Street community residents that their neighborhood was safe after the "clean-up" in 2001. But the concerned citizens of Agriculture Street Landfill disagreed and filed a class-action lawsuit against the city of New Orleans for damages and relocations costs. Unfortunately, it was Katrina that accomplished the relocation – albeit a forced one.

In January 2006, after 13 years of litigation, Seventh District Court Judge Nadine Ramsey ruled in favor of the residents – describing them as poor minority citizens who were "promised the American dream of first-time homeownership," though the dream "turned out to be a nightmare" (Finch, 2006, p. A1).

The case is currently on appeal. A year after the storm, a dozen or so FEMA trailers housed Katrina survivors in the contaminated neighbor-hood – where EPA announced in April 2006 it had found the carcinogen benzo(a)pyrene at levels almost 50 times the health screening level. No decision has been made by EPA to clean up the contamination found near the old Agriculture Street landfill (Brown, 2006).

A year after the storm, tons of trash, hurricane debris, flooded cars, and contents from gutted homes and businesses still line some neighborhoods streets. The Army Corps of Engineers is the agency charged with one of the biggest environmental cleanup ever attempted: scraping miles of sediment laced with cancer-causing chemicals from New Orleans' hurricane-flooded neighborhoods (Loftis, 2005).

Katrina floodwaters contained a "soup of pathogens" and contaminated muck (Cone, 2005; Cone & Powers, 2005; Dunn, 2005; CDC and EPA, 2005). Sediments of varying depths were left behind by receding Katrina floodwaters, primarily in areas of New Orleans impacted by levee overtopping and breaches (Horne, 2006). More than 100,000 of New Orleans's 180,000 houses were flooded, and half sat for days or weeks in more than 6 feet of water. Government officials estimate that as many as 30,000–50,000 homes citywide may have to be demolished, while many others could be saved with extensive repairs (Nossiter, 2005a, 2005b).

Although government officials insist the dirt in residents' yards is safe, Church Downs, Inc., the owners of New Orleans' Fair Grounds, felt it was not safe for its million-dollar thoroughbred horses to race on. The Fair Grounds is the nation's third-oldest track – only Saratoga and Pimlico have been racing longer. The owners hauled off soil tainted by

Hurricane Katrina's floodwaters and rebuilt a grandstand roof ripped off by the storm's wind (Martell, 2006). The Fair Grounds opened on Thanksgiving Day 2006. If tainted soil is not safe for horses, surely it is not safe for people – especially not for children who play and dig in the dirt.

Government failure to address post-disaster needs of African Americans has led to a "second disaster" (Bullard, 2005c). Not trusting government to respond to the needs of New Orleans African American communities, in March 2006, the Deep South Center for Environmental Justice (DSCEJ) and the United Steelworkers (USW), undertook "A Safe Way Back Home" initiative – a proactive pilot neighborhood clean-up project and the first of its kind in New Orleans (Gyan, 2006). The voluntary clean-up project, located in the 8,100 block of Aberdeen Road in New Orleans East, removed several inches of tainted soil from the front and backyards, replacing the soil with new sod, and disposing the contaminated dirt in a safe manner. Participants included residents and Steelworkers who have received training in hazardous materials handling programs funded by the NIEHS.

The broader goal of the "A Safe Way Back Home" was to provide a sustained effort over the next several months as hundreds of thousands of survivors of this disaster – many of whom are poor, disenfranchised, and African American – to begin the long, painful task of rebuilding their lives. Much of the work of this project focuses on the research, policy, and community outreach and assistance and education of the displaced minority population of New Orleans. The DSCEJ/USW coalition received dozens of requests and inquiries from New Orleans East homeowners associations to help clean up their neighborhoods block-by-block.

Eleven months after Hurricane Katrina struck, the federal EPA issued its final sediment report, giving New Orleans and surrounding communities a clean bill of health (U.S. EPA 2006). EPA deemed New Orleans safe. Government officials concluded that Katrina did not cause any appreciable contamination that was not already there. The agency pledged to monitor "pockets of contamination" and toxic "hot spots."

Although EPA tests confirmed widespread lead in the soil – a pre-storm problem in 40 percent of New Orleans – the agency dismissed residents' call to address contamination problem as outside of its mission. Federal and state officials see no need to scrape up the three million cubic yards of mud left by Katrina. The sole EPA recommendation for soil removal include soil near the million-gallon Murphy Oil spill in St. Bernard Parish and a 6-foot by 6-foot plot in Audubon Park – where lead contamination was found near a playground that did not flood (Brown, 2006). A broad coalition of scientists, health experts, environmentalists, and local residents view EPA's post-Katrina decision to simply monitor the contamination – rather than to clean it up – as a missed opportunity.

CONCLUSION

There is a racial divide in the way the U.S. government responds to natural and man-made disasters, such as hurricanes and floods, and public health threats (natural and man-made) in white and African Americans communities. This racial divide is not a recent phenomenon. For decades, black disaster victims have been treated as "second-class" citizens, discriminated in life and death.

Hurricane Katrina exposed the systematic weakness of the nation's emergency preparedness and response. Race impacts the speed and level of cleanup of toxic waste sites in the country. Clearly, race skews government response to emergencies, whether natural or man-made, with white communities seeing faster action and better results than communities where blacks, Hispanics, and other minorities live. What gets cleaned up and where the waste is disposed are key equity issues.

Dozens of toxic "time bombs" are still ticking away in communities where African Americans and other people of color are fighting against environmental racism, demanding equal protection of public health, relocation from toxic "hot spots," and equal treatment before and after disasters strike. They deserve no less. No Americans – black or white, rich or poor, young or old, sick or healthy – should have to endure needless suffering from a disaster.

NOTES

1. See Alabama Department of Archives & History. This Week in Alabama History – Compiled by Month. Available at http://www.archives.state.al.us/thisweek/month.html (accessed on October 17, 2005).

2. For a detailed description of Elba's floods, see Southern Regional Headquarters, National Weather Service, "Elba Alabama Levee Dedication 13 May 2002". Available at http://www.srh.noaa.gov/tlh/hydro/Elba_Levee_Dedication/Elba_Levee_Dedication.htm (accessed on October 10, 2005).

3. See Southern Regional Headquarters, National Weather Service, "Elba Alabama Levee Dedication 13 May 2002," for individuals who attended the dedication and groundbreaking ceremony. They include, Miss Elba, Kaye Whitworth, Elba Chamber of Commerce, Ferrin Cox, Chairman of the Governor's Long Range Task Force Committee of the Levee Project, Junior Miss Elba, Barbara

Everett (wife of U.S. Congressman Terry Everett), Governor Siegelman plus Future Little Miss Elba, U.S. Senator Richard Shelby plus Little Miss Elba, Alabama Speaker of the House, Seth Hammett, Elba Mayor, James Grimes, Alabama State Representative, Terry Spicer, Alabama State Senator, Jimmy Holley, Col. Robert Keyser, USACE Mobile District, Division Engineer, Todd Davison, FEMA , and Lee Helms, Director of Alabama Emergency Management Agency.
   4. Highlights of the workshop on environmental justice and emergency response were presented at the May 2005 Conference of Black Trade Unionists (CBTU) Community Action and Response Against Toxics (CARAT) Team Annual Conference in Phoenix, Arizona. Available at http://www.broadcasturban.net/webcast/cbtu2005/sat_workshop.htm
   5. The *People of Color Environmental Groups Directory* was funded by small grants from the Charles Stewart Mott Foundation and the Ford Foundation.

REFERENCES

Associated Press. (2007). Homeowners win Katrina damages case, *CBS News*, January 11. Available at http://www.cbsnews.com/stories/2007/01/11/katrina/main2351848.shtml

Bolin, R., & Bolton, P. A. (1986). *Race, religion, and ethnicity in disaster recovery*. Boulder, CO: Institute of Behavioral Science, University of Colorado.

Brinkley, D. (2006). *The great deluge: Hurricane Katrina, New Orleans, and the Mississippi Gulf Coast*. New York: William Morrow.

Brown, M. (2006). Final EPA report deems N.O. safe. *The Times Picayune*, (August 19).

Bullard, R. D. (1983). Solid waste sites and the black Houston community. *Sociological Inquiry*, 53(Spring), 273–288.

Bullard, R. D. (1987). *Invisible Houston: The black experience in boom and bust*. College Station, TX: Texas A&M University Press.

Bullard, R. D. (1994). Unequal environmental protection: Incorporating environmental justice in decision making. In: A. M. Finkel & D. Golding (Eds), *Worst things first? The debate over risk-based national environmental priorities* (pp. 237–266). Washington, DC: Resources for the Future.

Bullard, R. D. (2000). *Dumping in Dixie: Race, class, and environmental quality* (3rd ed.). Boulder, CO: Westview Press.

Bullard, R. D. (2005a). *The quest for environmental justice: Human rights and the politics of pollution*. San Francisco: Sierra Club Books.

Bullard, R. D. (2005b). EPA's draft environmental justice strategic plan – A giant step backward, July 15. Available at http://www.ejrc.cau.edu/BullardDraftEJStrat.html

Bullard, R. D. (2005c). Katrina and the second disaster: A twenty-point plan to destroy black New Orleans, December 23. Available at http://www.ejrc.cau.edu/Bullard20PointPlan.html

Business Week. (2005). The mother of all toxic cleanups, September 26. Available at http://www.businessweek.com/magazine/content/05_39/b3952055.html

Callimachi, R., & Bass, F. (2006). AP: Whites on top of Katrina Insurance: Whites pursued Katrina Insurance complaints more aggressively than minorities. *CBS News*, October 24. Available at http://www.cbsnews.com/stories/2006/10/24/ap/national/mainD8KVA2SG0.shtml

Centers for Disease Control and Prevention and EPA. (2005). Environmental health needs and
    habitability assessment. In: *Atlanta: Joint task force Hurricane Katrina response, initial
    assessment*, September 17. Found at http://www.bt.cdc.gov/disasters/hurricanes/katrina/
    pdf/envassessment.pdf#search = 'centers%20for%20disease%20control%20katrina%20
    contamination'.
Clinton, W. J. (1994). Federal actions to address environmental justice in minority populations
    and low-income populations, Exec. Order No. 12898. *Federal Register*, *59*(32), February
    11. Available at http://www.epa.gov/compliance/resources/policies/ej/exec_order_12898.pdf
    #search = 'executive%20order%2012898'
Commission for Racial Justice. (1987). *Toxic wastes and race in the United States*. New York:
    United Church of Christ.
Cone, M. (2005). Floodwaters a soup of pathogens, EPA finds. *Los Angeles Times* (September 8),
    A18.
Cone, M., & Powers, A. (2005). EPA warns muck left by floodwaters is highly contaminated.
    Los Angeles Times, (September 16).
Council on Environmental Quality. (1997). *Environmental justice: Guidance under the national
    environmental policy act*. Washington, DC: December 10.
Cuevas, F. (2005). Fla eyes strengthening Wilma. *The Atlanta Journal – Constitution* (October 18),
    A6.
Dunn, M. (2005). Sampling shows contamination. *The Advocate*, (October 7).
Dyson, M. E. (2006). *Come hell or high water: Hurricane Katrina and the color of disaster*.
    New York: Basic Books.
Federal Emergency Management Agency. (1999). *Hazard mitigation at work: Two Alabama
    communities*. April 15: FEMA Region IV.
Federal Emergency Management Agency. (2006). By the numbers – one year later: FEMA
    recovery update for Hurricanes Katrina, Press Release, August 22. Available at http://
    www.fema.gov/news/newsrelease.fema?id = 29109 (accessed on September 19, 2006)
Finch, S. (2006). Ag street landfill case gets ruling: City ordered to pay residents of toxic site.
    The Times-Picayune, (January 27).
Griggs, T. (2005). Rebuilding to be slow, expensive. *The Advocate* (September 11), 12A.
Gyan, J., Jr. (2006). Project brings green to N.O. *The Advocate*, (March 24).
Hartman, C., & Squires, G. (2006). *There is no such thing as a natural disaster*. New York:
    Routledge.
Horne, J. (2006). *Breach of faith: Hurricane Katrina and the near death of a great American city*
    (1st ed.). New York: Random House.
Kleinberg, E. (2003). *Black cloud: The deadly Hurricane of 1928*. New York: Carroll & Graf
    Publishing.
Lareau, A. (2000). *Home advantage: Social class and parental intervention in elementary
    education* (2nd ed.). Lanham, MD: Rowman & Littlefield.
Lee, A. (2005). Wind or water: The debate rages, but who will pay? *The Sun Herald (South
    Mississippi)*, (December 21).
Loftis, R. L. (2005). Extreme cleanup on tap in New Orleans. *The Dallas Morning News*,
    November 6. Available at http://www.dallasnews.com/sharedcontent/dws/dn/
    latestnews/stories/110605dntswtoxic.c3d4a5d.html
Louisiana Department of Environmental Quality. (2005). Hurricane Katrina Debris Manage-
    ment Plan, September 28. (revised on September 30), Appendix A.

Lyttle, A. (2004). Agricultural street landfill environmental justice case study. University of Michigan School of Natural Resource and Environment. Found at http://www.umich. edu/~snre492/Jones/agstreet.htm (accessed on October 6, 2004).

Martell, B. (2006). Horse racing returns to New Orleans. *Associated Press*, (November 23).

National Fair Housing Alliance. (2005). *No home for the holidays: Report on housing discrimination against Hurricane Katrina survivors – Executive Summary*. Washington, DC: NFHA, December 20. Available at http://www.nationalfairhousing.org/html/ Press%20Releases/Katrina/Hurricane%20Katrina%20Survivors%20-%20Report.pdf

Nossiter, A. (2005a). Many New Orleans homes doomed. *The Atlanta Journal – Constitution* (October 23), A8.

Nossiter, A. (2005b). Thousands of demolitions are likely in New Orleans. *The New York Times*, October 23. Available at http://www.nytimes.com/2005/10/23/national/national-special/23demolish.html?ex = 1287720000&en = bffeb9e9c0b34472&ei = 5088&part-ner = rssnyt&emc = rss.

Pastor, M., Jr., Bullard, R. D., Boyce, J., Fotehrgill, A., Morello-Frosch, R., & Wright, B. (2006). *In the wake of the storm: Environment, disaster and race after Katrina*. New York: The Russell Sage Foundation.

Roberts, J. T., & Toffolon-Weiss, M. (2001). *Chronicles from the environmental justice frontline*. New York: Cambridge University Press.

Saulny, S. (2006). 5,000 public housing units in New Orleans are to be razed. *The New York Times*, (June 15).

Tanneeru, M. (2005). It's official: 2005 Hurricanes blew records away. *CNN.Com*, December 30. Available at http://www.cnn.com/2005/WEATHER/12/19/hurricane.season.ender/

Uchitelle, L. (2005). Jobs surged last month in rebound from storm. *The New York Times*, (December 3).

U.S. Environmental Protection Agency. (1998). *Guidance for incorporating environmental justice in EPA's NEPA compliance analysis*. Washington, DC: U.S. EPA.

U.S. Environmental Protection Agency. (2006). *Summary results of sediment sampling conducted by the Environmental Protection Agency in response to Hurricanes Katrina and Rita*. EPA website, August 17. Available at http://www.epa.gov/katrina/testresults/sediments/ summary.html

U.S. General Accountability Office. (1983). *Siting of hazardous waste landfills and their correlation with racial and economic status of surrounding communities*. Washington, DC: U.S. Government Printing Office.

Van Heerden, I. (2006). *The storm: What went wrong and why during Hurricane Katrina*. New York: Viking.

Wright, B. (2005). Living and dying in Louisiana's Cancer Alley. In: R. D. Bullard (Ed.), *The quest for environmental justice: Human rights and the politics of pollution*. San Francisco: Sierra Club Books.

Wright, B. H., & Bullard, R. D. (2007a). Washed away by Hurricane Katrina: Rebuilding a 'new' New Orleans. In: R. D. Bullard (Ed.), *Growing smarter: Achieving livable communities, environmental justice, and regional equity* (pp. 198–212). Cambridge, MA: MIT Press.

Wright, B. H., & Bullard, R. D. (2007b). Wrong complexion for protection: Will the 'mother of all toxic cleanups' in post-Katrina New Orleans be fair. In: R. D. Bullard, P. Mohai, R. Saha & B. H. Wright (Eds), *Toxic wastes and race at twenty, 1987–2007: Grassroots struggles to dismantle environmental racism* (pp. 124–133). Cleveland, OH: United Church of Christ.

PART II:
RACE AND GENDER IN ENVIRONMENTAL MANAGEMENT AND ANALYSIS

DIVERSITY AND THE ENVIRONMENT: MYTH-MAKING AND THE STATUS OF MINORITIES IN THE FIELD

Dorceta E. Taylor

ABSTRACT

There is growing interest in diversity in the environmental field. The issue has become more pertinent as country undergoes noticeable demographic changes. Researchers have been interested in diversity for sometime too. This chapter traces the evolution of research on diversity and the environment. It discusses the results of new studies examining students' attitudes toward their work in environmental organizations as well as their salary expectations. The chapter also analyzes the demographic characteristics of the leadership of environmental institutions as well as their hiring and recruiting practices.

INTRODUCTION

I was flabbergasted by his response. It was a simple statement, but one that changed the course of my career. The year was 1981, and I signed up for an

Equity and the Environment
Research in Social Problems and Public Policy, Volume 15, 89–147
Copyright © 2008 by Elsevier Ltd.
ISSN: 0196-1152/doi:10.1016/S0196-1152(07)15003-1

environmental studies course reluctantly because both the advanced zoology and botany courses I wanted to take were oversubscribed. Until that point in my undergraduate studies, I had taken biology and chemistry courses and they always had a few blacks and other students of color in them – most of those minority students were premed. In my case, I was preparing for a career as a geneticist.

I was taken aback the first time I walked into the environmental course – everyone except me was white. There were about 60 students in the class and the only seats available were in the back of the room. I enjoyed the subject matter but noticed that the teacher spoke to and interacted with only the students in front. As a former high school science teacher, I knew it was possible to and desirable for a teacher to engage all the students, so I decided to keep my seat in the back of the room and remain quiet with all the other back-benchers to see how long it would take the professor to notice that there were students there too. This went on until the first exam – almost half way through the semester. After announcing that he was returning the exams with the highest scores first, he handed out all the graded exams except mine. When he inquired who Dorceta was and I raised my hand, a look of veiled surprise flitted across his face momentarily. He asked me to pick up my exam after class. Although I had taken many science courses, none had integrated the social and human dimensions of environmental problems to the extent this course did – and for me, therein lay the intrigue. Thinking that students of color could benefit from the course and use it to help their communities, I decided to raise the topic with the teacher when I collected my exam. It turned out I had the highest score in the class (not the lowest, as students were led to believe from the order in which the exams were returned) and the teacher wanted to know who I was and why I did so well on the exam. As I turned to leave, I asked what I thought was an innocuous question. "Why aren't there more black students in this course?" I queried. The next words he uttered shocked me. Without giving it a moment's thought, he said in a matter-of-fact voice, "Because blacks are not interested in the environment." I did not respond. I could not. His answer was so unexpected that words escaped me.

My interest in the life sciences was awakened and nurtured in Jamaica. While in school there, I spent countless hours indulging my obsession with the natural and built environment. I was by no means unique in the Caribbean (or in Africa, Asia, or Latin America for that matter). At any given time, one can find thousands of black, Indian, Chinese, and Korean students, like myself, trekking long distances in suffocating heat to collect specimens; studying the flora and fauna of their countries; investigating marine ecosystems; and studying landforms, pollution, and erosion, etc. My experience in America also told me there were other blacks like myself interested and intrigued by the environment. However, until that fateful fall evening, it had not occurred to me that environment was something that whites were interested in, concerned about, and was knowledgeable about, and blacks and other minorities were not (or were not supposed to be). Nothing in my experience indicated that this was the case or prepared me to accept it as a fact. Shortly after the conversation with the professor, I went to the library to find out what the scholarly literature had to say about the topic. Astonishment gave way to incredulity when I realized that books and journal articles seemed to support the professor's statement. The die was cast. Since then, I have been studying diversity as well as the historical and contemporary status of minorities in the environmental field.

Despite decades of research, debate and activism, diversity is still a salient topic in the environmental movement. Scholars and practitioners are interested in knowing about the state of diversity in environmental institutions, the participation of minorities in environmental affairs, and equity (as it relates to benefits, opportunities, and negative impacts). Environmentalists are also concerned with perceptions, attitudes and behaviors, and the relationship between these factors and the framing of discourses, as well as activism. This essay will examine how the research and discourse on diversity in the environmental field has evolved over the past four and a half decades. It will also present data from new studies conducted by this author that examine some pervasive myths and questions related to minorities and the environment.

THEORETICAL FRAMEWORK

This chapter examines the evolution of the discourse on diversity by drawing on theoretical concepts from the social movements literature on framing; race relations literature on prejudice, discrimination, and racism; and network and organizational literature on filtering, homophily and homo-social reproduction. This framework will help the reader understand the evolution of the diversity discourse. It will also help further our understanding of the current status of diversity in the environmental field and point to mechanisms for changing conditions.

Framing

Framing has played a key role in the discourse on diversity and in the perception of how minorities relate to the environment. Environmental activists, policy makers, government, politicians, and businesses have long perceived, contextualized and battled over environmental issues by establishing frames of reference. According to Snow and Benford (1992) and Gamson (1992), framing is the process by which individuals and groups identify, interpret, and express social and political grievances. A central feature of the framing process is the generation of diagnostic attributions; that is, the identification of problems and the imputation of blame or causality. The environmental justice frame (Taylor, 2000; Capek, 1993) is an emergent environmental frame arising from the environmental activism of people of color in the U.S. The frame bridges ideologically compatible but structurally separate issues by explicitly linking environmental experiences with racial, class, gender, and other forms of social inequality. It links ethnic minorities and the poor with the increased possibility of exposure to environmental hazards and decreased likelihood of exposure to positive environmental amenities.

Effective frames are potent, i.e., they resonate with their target audience. The extent to which a frame resonates is influenced by three interrelated factors: empirical credibility, experiential commensurability, and ideational centrality or narrative fidelity. Thus there has to be credible empirical evidence to support the frame's claims, the target audience has to have some experience with the problem, and the issue must strike a chord in people. In other words, the frame must salient enough for potential recruits to identify with it (Snow & Benford, 1988; Fisher, 1984). Depending on what stage the social movement is in, movement activists must balance the use of the rhetoric that appeal to morals or values with and efforts to ratify claims through evidentiary support or by proposing policy alternatives (Best, 1987; Gamson & Meyer, 1996; Rafter, 1992; Ibarra & Kitsuse, 1993; Hannigan, 1995).

Prejudice, Discrimination and Racism

As later discussion shows, early on in the discussion of diversity, people of color were portrayed or framed in negative terms vis-à-vis the environment. Negative or inaccurate framing can lead to stereotyping, and stereotyping is related to prejudice and discrimination. Although the terms "prejudice" and "discrimination" are often used synonymously in every day language, the meanings differ. Prejudice is a set of beliefs and stereotypes that lead an individual or group of people to be biased for or against members of a particular group. Although prejudice can lead to discrimination, this does not always happen (Aguirre & Turner, 1998; Cashmore, 1988; Merton, 1949).

Even though prejudice may not necessarily lead to discrimination, it is an important factor that can trigger discrimination. As Aguirre and Turner (1998) argue, prejudicial beliefs and stereotypes often highlight – unfairly and inaccurately – characteristics of ethnic groups that focus on negative characteristics. The negative imagery can be used to legitimize discrimination. Discrimination refers to the unfavorable treatment of people assigned to a particular social group (Banton, 1988a). Prejudice also instills fear and anger; these emotions and feelings can trigger unprovoked hostility and acts of discrimination. Prejudice also creates an atmosphere of intolerance, which can manifest itself through acts of discrimination (Aguirre & Turner, 1998).

Racism

Up to the late 1960s, racism was commonly defined as a doctrine, dogma, ideology, or set of beliefs. The central theme of this doctrine was that race determined culture. Some cultures were deemed superior to others, therefore, some races were superior and others inferior. During the 1960s, the definition of racism was expanded to include the practices, attitudes, and beliefs that supported the notion of racial superiority and inferiority. Such beliefs and practices produced racial discrimination. In 1967, Stokley Carmichel and Charles Hamilton interjected the concept of "institutional racism" into the discourse. The term was used to describe the institutional processes and apparatus that support and maintain racial discrimination (Carmichel & Hamilton, 1967; Banton, 1988b; Banton & Miles, 1988).

Contemporary scholars argue that to limit the understanding of racism to prejudicial and discriminatory behavior misses important aspects of racism. Racism is also a system of advantages or privileges based on race. In the American context, many of the privileges and advantages that accrue to whites stem directly from racial discrimination directed at people of color. However, many whites do not think racism affects them. Neither do they consider whiteness as a racial identity that carries with it certain advantages and privileges (Dyer, 1988; Flax, 1993; Katz & Ivey, 1992). Because of this, Omi and Winant (1993) contend that whites have a transparent racial identity. That is, as a signifier of power, privilege, and dominance, whiteness often remains invisible to the bearer and user. As Peggy McIntosh (1990) argues, because whites are not taught to recognize white privilege as an aspect of racism, white privilege can be seen as an invisible package (or weightless knapsack) of special provisions, unearned assets and benefits that they can count on cashing in each day, but about which they remain oblivious. Therefore, racism results not only from personal ideology and behavior; those personal thoughts and actions are supported by a system of cultural messages and institutional policies and practices. Thus racism can be more fully described as the execution of prejudice and discrimination coupled with power and privilege (Brah, 1992; Lorber, 1994; Lucal, 1996; Wellman, 1977).

Filtering, Homophilous Networks, and Homosocial Reproduction

Workforces that lack diversity arise by a variety of means. In addition to factors such as stereotyping, prejudice, and discrimination, homogeneous workplaces arise because of filtering and other recruitment practices. Filtering occurs through the practice of hiring of staff from the same social and professional networks, i.e., hiring from the same firms or amongst people trained in the same institutions, by employing similar promotion practices (such as promoting top executives only from specific departments in an organization), or from the skill requirements for particular jobs. Universities and professional associations socialize individuals, thereby creating a pool of interchangeable, almost indistinguishable people who occupy similar positions in a range of organizations. These employees have similar orientations and dispositions that may override the variations in tradition and control that might otherwise foster diversity (DiMaggio & Powell, 1983, 1991; Perrow, 1974). Filtering promotes the reliance on homophilous networks, i.e., using practices like insider referrals to recruit from among networks of similar people. Reliance on such networks results in the homosocial reproduction of the workforce – a replication of the demographic and social characteristics of the existing workforce – because people tend to refer others similar to themselves for jobs. Such practices allow race to play a role in recruitment and hiring practices (DiMaggio & Powell, 1991; Granovetter, 1995; Braddock & McPartland, 1987; Model, 1993).

DIVERSITY, EQUITY, AND THE ENVIRONMENT: THE INKLINGS OF AN IDEA

Although minorities and the poor are often portrayed as being unconcerned about and unaware of the environment, this assumption bears questioning. For instance, letters unearthed in New York City's municipal archives show that, as early as the 1700s, the general public was concerned about the environment and wanted to see improvements. In 1797, for instance, residents of the Sixth Ward – one of the city's poorest neighborhoods – complained to the mayor about a stagnant pond and the "number of dead animals being thrown into it." According to the letter, the pond was "now in a state of putrefaction, together with a pernicious matter running from a glue manufactory, caus[ing] your petitioners to be apprehensive that if left as is at present during the hot season it may prove fatal to the health of the inhabitants that live near the same." On August 21, 1888, Albert Oelzer of Henry Street wrote to complain about a dead horse in the street that had not been carted away. Oelzer wrote "The stench is unbearable, and people in the neighborhood of which I am one were forced to sleep with closed windows last night. Not a pleasant thing, I assure you" (Municipal Archives, 1797, 1888).

Despite these expressions of concern from ordinary citizens, a mass environmental movement did not emerge until the second half of the 20th century. However, during the late 18th and early 19th centuries, a small cadre of environmental activists emerged to spearhead environmental initiatives and form the earliest environmental organizations. These activists were white and primarily male. They were also upper-class sportsmen, lawyers, politicians, policymakers, businessmen, and other civic leaders (Taylor, forthcoming; Warren, 1997; Judd, 1997). Throughout the history of the movement, elites have played prominent roles in it. This has sometimes resulted in a failure to recognize the role of ordinary citizens in environmental affairs. This is particularly true of minorities.

For the first century of its existence, the environmental movement achieved many remarkable feats as it gained power and prestige. Although the movement grew steadily, it grew slowly through the first half of the 20th century. The great transformation of the movement occurred during the 1960s and 1970s (Taylor, 1999). About the same time that the movement was undergoing rapid changes and a mass movement was emerging in the 1960s, scholars began examining its demographic characteristics. They also began to ask about the role of blacks and other minorities in the movement.

However, the 1960s was not the first time questions were raised about the composition of the environmental movement. During the 19th century – when the movement was in its infancy – working-class activists and recent immigrants noted the middle- and upper-class membership of the movement and environmental organizations and raised questions about inequalities arising from the promulgation of some environmental laws. As the 19th century drew to a close, upper-class women also began to question the nature of the environmental discourse that seemed to target women as the source of bird destruction while seeming to absolve men from much of their destructive behavior. While the women responded to perceived inequities in the environmental discourse by taking on leadership roles in environmental organizations and becoming more active in environmental affairs, the working class – generally excluded from membership in these organizations – responded by flouting environmental regulations or challenging them in the courts. By and large, the working class continued to express their dissatisfaction as challengers and outsiders of the fledgling conservation movement (Taylor, forthcoming; Warren, 1997; Judd, 1997; Doughty, 1975).

THE DEMOGRAPHIC CHARACTERISTICS OF THE MEMBERSHIP OF ENVIRONMENTAL GROUPS

The first wave of research on diversity in the environmental movement began during the 1960s. As the contemporary environmental movement took shape in that decade, scholars and social observers raised questions about the racial, class, and gender composition of the movement. These early studies approached the topic of diversity by examining the demographic composition of the membership and leadership of environmental groups.

Analyses of the membership of environmental organizations generally found a middle-class membership. For instance, a 1969 national survey of 907 Sierra Club members indicated that the organization had a middle-class membership. At a time when 11% of the general public had a college degree, 74% of Sierra Club members had at least a college degree, and 39% had an advanced degree (Devall, 1970; Census, 2000a). A 1971 study of the Puget Sound chapter of the Sierra Club also found a very similar profile. Ninety-seven percent of the membership had attended college, 88% had a bachelor's degree, and 71% had an advanced degree (Faich & Gale, 1971). Similarly, a 1972 study of 1,500 environmental volunteers nationwide showed that 98% of them were white and 59% held a college or graduate degree (Zinger, Dalsemer, & Magargle, 1972). In 1980 and 1982, Milbrath (1984) conducted studies in which he compared 225 and 274 environmentalists respectively with the general population. The studies found environmentalists were more highly educated and more likely to hold professional jobs than the general population. The studies also found that while the general population was 83% white, the 1980 environmental sample was 92% white; so was 94% of the 1982 environmental sample.

Other studies in this genre have discussed the middle-class nature of the membership of the environmental movement at length (Buttel & Flinn, 1974, 1978; Cotgrove & Duff, 1980; Dillman & Christenson, 1972; Harry, 1971; Harry, Gale, & Hendee, 1969; Hendee, Gale, & Harry, 1969; Lowe, Pinhey, & Grimes, 1980; Tognacci, Weigel, Wideen, & Vernon, 1972). Some found education, income, and race to be associated with naturalistic values and environmental concern (Harry, 1971; Hendee et al., 1969; Dillman & Christenson, 1972; Tognacci et al., 1972; Wright, 1975; Martinson & Wilkening, 1975; Kellert, 1984). Because these studies did not examine barriers to participation or the alternative forms of environmental activities minorities and the poor participated in, minorities were often portrayed as disinterested, disengaged, or under-participating in environmental affairs.

MINORITIES: ATTITUDES, PERCEPTIONS, AND PARTICIPATION IN ENVIRONMENTAL AFFAIRS

In the same way questions about the demographic composition of environmental organizations generated research, a parallel body of scholarship emerged on the environmental attitudes of minorities and their participation in environmental affairs. In particular, scholars focused on awareness of environmental issues, concern for the environment, membership in environmental organizations, and participation in environmental activities. These studies also focused on minority participation in mainstream environmental activities (like bird watching, hiking, mountaineering, interest in wildlife, etc.). Scholarly inquiry all but ignored the environmental activism (related to civil rights, Indian rights, Chicanismo, etc.) going on in minority communities, as well as the environmental activities that minorities were participating in. Not surprisingly, these studies found that minorities were less likely to participate in environmental affairs, and that they showed less knowledge and awareness of and concern for environmental issues than whites. In explaining their results, researchers argued that minorities were too concerned with meeting their basic needs to focus on environmental issues (van Ardsol, Sabagh, & Alexander, 1965; Kreger, 1973; Meeker, Woods, & Lucas, 1973; Van Liere & Dunlap, 1980; Morrison, Hornback, & Warner, 1972; Kellert, 1984).

Researchers interested in the topic saw access to environmental amenities as another aspect of diversity worth investigating. Consequently, they examined minority participation in outdoor recreational pursuits. As was the case with other genres of early diversity studies, investigators explored participation in activities typically pursued by the white middle class and paid less attention to those generally participated in by minorities and the poor. Predictably, the results indicated that blacks and other minorities had lower rates of participation in outdoor activities such as camping, wilderness, or national park visits than whites. Studies also found that minorities tended to recreate closer to home, using city parks and other local recreational facilities rather than distant national, state, or regional recreational sites (Mueller & Gurin, 1962; Washburne, 1978; Washburne & Wall, 1979, 1980; Yancey & Snell, 1971; Craig, 1972; O'Leary & Benjamin, 1982; Hartmann & Overdevest, 1989; Stamps & Stamps, 1985; Noe, 1974; Burdge, 1969; Kelly, 1980; Dwyer & Hutchison, 1988; Kornegay & Warren, 1969; Enosh, Staniforth, & Cooper, 1975). In general, these studies helped to reinforce the perception that minorities were disinterested in the environment.

RECASTING THE DEBATE: RETHINKING THE RESEARCH ON MINORITIES AND THE ENVIRONMENT

During the 1980s, a group of scholars began to analyze and critique the research on minorities and the environment. At the same time, they began to put forth new models and ideas that changed the way we understood the relationship between minorities and the environment. This body of scholarship not only cast doubt on the interpretations of minorities' environmental attitudes and behaviors that had been reported in earlier studies, but also the perceptions of minorities that these studies had fostered. The new wave of scholarship argued that minorities were interested in the environment, were concerned about environmental issues and were participating in environmental affairs. This wave of scholarship also pointed out how biases in the types of environmental activities studied could lead to the erroneous conclusion that minorities were disinterested and under-participating in environmental affairs. These studies also pointed to institutional and other barriers that affected minority environmental participation (Mohai, 1985, 2003; Mohai & Bryant, 1998; Taylor, 1989, 1992). This author was among the scholars who began to question the findings and explanations of earlier studies. Her first publication on this topic stemmed from the research begun in the fall of 1981, soon after her encounter with the environmental studies professor.

USHERING IN THE CONTEMPORARY DIVERSITY DISCOURSE: A LETTER AND SOME CONFERENCES

January 1990 was a watershed moment in the on-going debate over diversity and the environmental movement. Two events occurred around that time that ushered in the modern discourse on diversity in the environmental field. On January 16th, several environmental justice activists sent a letter to the heads of the country's largest environmental non-governmental organizations (NGOs)[1] claiming that the organizations hired few minorities and were alienated from minority communities where pollution and other environmental ills were prevalent. The letter, published in the *New York Times* on February 1st, forced mainstream environmental activists – who were working feverishly on preparations for the 20th anniversary celebrations of Earth Day – to stop and explain to curious reporters why so few minorities were hired in environmental organizations (Shabecoff, 1990).

Representatives of the targeted organizations responded to the letter by saying that, although the organizations had a poor track record of hiring minorities, racism was not a factor in their hiring decisions. The spokespersons also hastened to say that they were trying to rectify the situation. At the time the letter was written, only 1.9% of the staff of the Audubon Society, Friends of the Earth, Natural Resources Defense Council, and Sierra Club were minorities. Echoing the reasoning of academics who had researched minority participation in environmental affairs during the 1960s and 1970s, leaders of the environmental organizations offered several explanations for the lack of minorities on their staff. They argued that minorities were not applying for jobs in environmental NGOs, the environmental groups did not recruit minorities aggressively, there was a scarcity of minorities among the pool of trained environmental specialists, and that minorities did not want to work for the low salaries being offered by environmental organizations. Furthermore, Frederic Krupp, executive director of the Environmental Defense Fund, argued that minorities are "cause oriented," being attracted to issues such as discrimination and poverty, rather than to environmental issues (Shabecoff, 1990).

The timing of this letter was significant. It appeared in the *New York Times* about three months before Earth Day 1990. At the time, plans were underway for a huge celebration of the achievements of the environmental movement. The letter – the first to question the hiring practices of the environmental groups openly – issued an ultimatum for increased hiring of minorities, and framed the demographic characteristics of the environmental workforce in terms of racist hiring practices (rather than lack of interest or inaction on the part of minorities). In short, the letter generated a lot of attention. The letter also coincided with other environmental justice organizing events, including the University of Michigan's conference on race and the incidence of environmental hazards, and the Agency for Toxic Substances and Disease Registry's National Minority Health Conference. The University of Michigan's conference was also significant in that it examined the relationship between race and the exposure to environmental hazards. The conference was attended by minority scholars and policy makers as well as progressive white colleagues. In addition to highlighting the problems of racial disparities in the exposure to environmental hazards, the Michigan conference focused on several other pressing issues. Conferees focused on the lack of diversity in both environmental NGOs and government environmental agencies, the lack of funding for minority environmental activists to do work in the field, and lack of effective policies to combat environmental inequalities (Bryant & Mohai, 1992; Institute of Medicine, 1999; Taylor, 2000).

The issue of diversity took center stage again at the 1991 First National People of Color Environmental Leadership Summit. Since then, diversity has remained a part of the environmental discourse. There have been a host of conferences, workshops, and other events pressing for increased diversity in environmental organizations.

PRIOR ENVIRONMENTAL JUSTICE ACTIVISM AND THE CHALLENGE TO MAINSTREAM ENVIRONMENTALISTS

The letter to the Green Group did not come from out of the blue. Native Americans, African Americans, Hispanics, and Asian Americans had been active in environmental issues for a long time. Moreover, they intensified their activism around issues like sovereignty, land rights, fishing and hunting rights, reduction in the use of pesticides, occupational health and safety, housing, lead poisoning, access to open space, solid waste disposal, and transportation equity issues during the civil rights, red power and Chicano movements of the 1960s (Taylor, 1997, 2002). Despite the heightened focus on environmental issues in minority communities, most environmentalists and scholars missed or ignored this activism because it did not have the traditional environmental framing, and it did not occur in traditional environmental settings or under the aegis of traditional environmental groups.

Consequently, minority environmental activists continued to organize and work completely under the radar screen of environmentalists – and those purporting to study them – well into the next two decades. By the 1970s, toxics campaigns moved to the forefront of minority environmental struggles. For instance, in 1978, residents of Triana, Alabama began fighting against dichlorodiphenyltrichloroethane (DDT) contamination (Harris, 1983; Maynard, Cooper, & Gale, 1995; Press, 1981; *Washington Post*, 1980). Similarly, in 1982, residents of Warren County (North Carolina) began protesting the construction of a hazardous waste landfill near Warrenton, in which the state planned to bury 400,000 cubic yards of soil contaminated with polychlorinated biphenyls (PCBs). Activists from the Warren County protests urged the General Accounting Office (USGAO) to examine the relationship between the location of landfills in the Southeast and the demographic characteristics of surrounding areas. This led to the publication of the 1983 USGAO study. Four years later, the United Church of Christ (UCC) Commission for Racial Justice (some of whose members had participated in the Warren County protests) published a national study examining the siting of hazardous facilities and waste sites. Both studies linked race and class with the increased likelihood of living close to hazardous facilities and toxic waste sites (*Washington Post*, 1982; LaBalme, 1988; U.S.G.A.O., 1983; UCC, 1987). The UCC study claimed that race was the most reliable predictor of residence near hazardous waste sites in the U.S. (UCC, 1987). This widely publicized study was the first to effectively bridge the race relations and environmental discourses in the United States.[2] The leaders of these and other on-going grassroots environmental justice campaigns were the ones who wrote the letter to the Green Group.

DIVERSITY IN THE ENVIRONMENTAL FIELD: DEMOGRAPHIC CHARACTERISTICS AND INSTITUTIONAL BARRIERS

The events of 1990 and 1991 ushered in a new wave of diversity studies. During the 1990s, there was also a veritable explosion in the number of environmental justice groups; hundreds of these groups were founded and led by minorities (Taylor, 1999, 2000). There was so much growth in this sector that by the end of the decade, it was no longer tenable to characterize minorities as disengaged from environmental affairs.

The diversity studies of this period were strongly influenced by the environmental justice discourse. They focused less on the lack of agency of minorities and more on the institutional factors that affected diversity. Consequently they focused on identifying institutional characteristics that hindered or promoted workforce diversity. They also examined the demographic characteristics of the staff, boards, and volunteers of these organizations. Researchers also began paying more attention to the role of the pipeline in diversity efforts. That is, investigators examined the extent to which minority students were pursing careers in the environmental field and how that was likely to affect workforce diversity in environmental organizations. By the 1990s, a new generation of leisure studies also examined minority leisure pursuits in a more nuanced manner. Not only did scholars begin to go beyond comparisons of blacks and whites, they also began to develop studies that helped them to understand minority participation better (Taylor, 1991; Ffloyd, 1998; Shaull & Gramann, 1998; Bowker & Leeworthy, 1998; Johnson et al., 1998; Shinew, Floyd, McGuire, & Noe, 1996; Yu & Berryman, 1996; McDonald & McAvoy, 1997).

Still, studies of the demographic characteristics of the staff of environmental organizations found them to be predominantly white. A 1990 poll of four of the largest environmental nonprofits found that only 14 (1.9%) of the 745 workers of the Audubon Society, Friends of the Earth, Natural Resources Defense Council, and Sierra Club were minorities (*New York Times*, 1990). Two years later, a Conservation Fund study of 265 leaders (president, chair, chief executive officer, etc.) of environmental organizations nationwide found that 79% of them were male and 3% were under 30 years old. Ninety-nine percent of the leaders had at least a bachelor's degree and one-fifth had a doctorate or other professional degree. Fourteen percent earned $60,000 or more annually. The Conservation Fund also studied 180 environmental volunteers nationwide in 1988. The study found that 61% of the volunteers were male, and only 7% were younger than 35 years of age. Seventy-nine percent had a bachelor's degree and 53% had an advanced degree. Seventy-one percent had a professional or managerial job, while only 3% worked as skilled laborers (Snow, 1992).[3]

A 1992 study of activists involved in disputes over the use of natural resources in the state of Washington found that environmentalists tended to be more educated and had higher incomes than labor activists. The environmental activists had an average annual income of $67,300, while forest industry workers averaged $29,300 and construction workers earned an average of $29,000. While 92% of the environmental activists had a college degree and 48% had advanced degrees, college degrees were much less common among timber and construction workers. Sixty-two percent of the environmental activists had professional occupations (Rose, 2000; DOI, 1992). In comparison, in 1992, the mean income for the general population was $23,227, and 21.4% of the population had four or more years of college (Census, 1990, 2000b, 2000c).

Another significant study was also published in 1992. The Environmental Careers Organization (ECO) study of 63 mainstream environmental organizations found that 32% of them had no people of color on their staff, 19% had no volunteers who were people of color, 22% had no board members who were people of color, and 16% had no people of color in their membership. Nonetheless, more than half of the organizations indicated that diversifying the organizations was a high priority in their organizations; most organizations wanted to diversify so that they could work better with communities of color. The study also found that 40% of the organizations had changed their recruiting practices to attract more minorities; 17% had established affirmative action plans, and 14% had began networking in minority communities. For the most part, the environmental organizations felt they did not have many people of color on staff because minorities did not apply for environmental jobs, the salaries were too low, or that there was a paucity of qualified minority applicants. Furthermore, most environmental organizations did not see any issue in their organization that would hamper the retention of minority workers; however, some thought low salaries would be a problem. Only 10% or less of the organizations thought the organizational culture, lack of awareness of minority issues, lack of group support for minority workers, and the inability of organizations to make minorities feel comfortable would make it difficult to retain people of color (ECO, 1992).

In the mid-1990s, not only did environmental organizations still attract a large following of males, but men still dominated the top leadership positions. This author's 1994 study of 1,467 environmental organizations found that, of the 1,402 organizations that listed the names of their leaders, 83% of the organizations had a male president, chair, executive director, or chief executive officer. Although women were more likely to be found in auxiliary leadership positions (such as secretary, librarian, magazine editor, program officer), they still occupied only 27% of the general leadership positions. In contrast, women were more likely to take on top leadership roles in environmental justice organizations. The author's 1994 study of 330 environmental justice groups found that, overall, 51% of the presidents, chairs, or executive directors were female (Taylor, 1999).

MINORITY STUDENTS: THE PIPELINE, ENVIRONMENTAL WORKFORCE PARTICIPATION AND SALARY EXPECTATIONS

Investigators also began to conduct pipeline studies that included minority students in the samples in the early 1990s. In contrast to the prevailing perception of minorities as unqualified, too few to be found, unwilling to seek jobs in environmental organizations, and unwilling to work for the salaries offered in such organizations, a 1991 ECO study (published in 1992) found a sizeable pool of minority students with appropriate backgrounds to work in the environmental field. At the time, about 13% of the students nationwide graduating with bachelor's degrees in the environmental disciplines were minorities; so were 11% of those graduating with a master's degree in the field. Moreover, the ECO student study found that minority students expressed a strong desire to work in nonprofits and government. Fifty-nine percent of the minority students indicated they would work in a national nonprofit and 65% would work in a grassroots nonprofit after graduation. Furthermore, 64% indicated they were willing to work with a government agency upon graduation (ECO, 1992).

The study also found that minority students had salary expectations that were well within the range of that being paid by environmental nonprofits. This is evident when salaries being paid by environmental organizations were compared to the salary expectations of minority students. The Conservation Fund's survey of 248 leaders of environmental organizations found that 13% were not compensated at all, while 33% earned $30,000 or less per annum. Fifty-two percent of the leaders earned between $31,000 and $100,000 per year, and the remaining 2% earned more than $100,000 annually (Snow, 1992). In 1990, environmental organizations reported that their starting salaries tended to range from $14,000 to $25,000 (*New York Times*, 1990). In 1991, 54% of the minority students surveyed said they expected to earn $25,000 or less per year when they graduated (ECO, 1992).

THE SALIENCE OF DIVERSITY: MINORITY STUDENTS AND THE ENVIRONMENTAL WORKFORCE

While salaries might not have been an issue for environmental leaders to be concerned about, diversity should have been. All the students of color surveyed indicated that a diverse workforce was very important to their job satisfaction. The data showed that, while some white workers in environmental organizations did not necessarily see workplace diversity as a critical factor in their job satisfaction, all of the minority students surveyed in the 1991 ECO study reported that workforce diversity was very important to their job satisfaction. This being the case, minorities might be reluctant to seek jobs or remain in work environments that lack diversity or show little progress toward becoming more diverse (ECO, 1992).

WE WANT TO HIRE THEM, BUT ...

Despite the interest in diversity, little has been done in the way of studying the state of diversity in environmental institutions since the early 1990s. Between 1992 and 2001, the periodic reports by the National Science Foundation on enrollment in science and engineering (S&E) disciplines and minority participation in the S&E workforce provided one type of national-level data that could be used to gauge the status of diversity in the environmental sector (for instance see, National Science Foundation, 1996, 2000). Although there are signs that diversity is increasing in environmental organizations, a recent diversity study indicates that much more can be done to enhance diversity. The 2002 report that examined diversity in 61 organizations in the Natural Resources Council of America found that 11.5% of the 6,347 staff and 9.6% of the 1,324 board members of these organizations were minorities (Stanton, 2002). How does this compare with the general population? In 2000, minorities comprised 30% of the U.S. population. Hispanics constituted 12.5% of the population, blacks 12.3%, Asians 3.6%, and Native Americans 0.9% of the population (Census, 2000b).

Minority hiring in the environmental sector is not a trivial concern. Historically, ethnic minorities have been underrepresented in all aspects of the environmental field. This fact is significant because the environmental field has grown to become an influential employment, research, and policy-making sector of the society. It is estimated that over $250 billion is spent on environmental activities in the U.S. each year. In addition, the environmental sector employs more than 1.8 million people in over 150,000 environmental agencies, companies, non-profit organizations, and academic institutions. Nonetheless, low percentages of minorities are on the staffs of the approximately 10,000 environmental organizations or are among the estimated 600,000 people employed in federal and state environmental agencies. Furthermore, relatively few minorities are hired in a professional capacity in this sector. As the report *Land of Plenty* argues, a large pool of potential workers (minorities and women) remain isolated from the science, engineering, and technology workforce even as demand has grown for workers with science, engineering, and technology training (ECO, 1992, 2001; CCAWMSETD, 2000).

The environmental sector relies heavily on scientists and to a lesser extent engineers to make up its workforce. Thus an examination of the S&E labor force will give us an indication of the status of minorities within the environmental field. Between 1983 and 2003, enrollment in American higher education institutions rose from 12.6 million to 15.7 million students. Demographers predict that the college population will continue to increase, because the number of people in the population between the ages of 20–24 years is expected to rise until about 2015. As the number of students enrolled in institutions of higher education has increased, so too has the number of S&E degrees awarded. Despite some fluctuations, the number of S&E bachelor's and master's degrees awarded between 1983 and 2002 attained new highs of 415,600 and 99,200, respectively, in 2002. Consequently, in 2002, S&E degrees accounted for 32% of the bachelor's degrees awarded. Moreover, after a four-year decline, the number of doctorates awarded rose in 2003 (National Science Board, 2006).

How is the status of women and minorities in S&E programs affected? The data show that the number of women enrolled in S&E disciplines has increased steadily for some time now. Females have earned 50% or more of the S&E bachelor's degrees awarded since 2002. In 1983, females comprised 36% of the graduate students in S&E; they comprised 47% of the group in 2003. The percentage of underrepresented minority students (blacks, Hispanics, and Native Americans) enrolling in S&E fields has also increased. Underrepresented minorities comprised 6% of S&E students in 1983; in 2003 they constituted 11%. Over the same period the percent of Asian students in S&E disciplines climbed to 7%. The proportion of bachelor's degrees awarded to minorities in S&E fields has also risen in this time period. From 1983 to 2002, the proportion of S&E degrees awarded to Asians increased from 4% to 9%, and the proportion awarded to underrepresented minorities increased from 9% to 16%. There have been increases in the proportion of master's degrees awarded to women and minorities in this period, as well. The proportion of S&E master's degrees awarded to women increased from 31% to 44%. The proportion grew from 5% to 7% for Asians and 5% to 11% for underrepresented minorities. The findings are similar at the doctoral level, where women earned 45% of the S&E doctorates awarded in 2003. The proportion of S&E doctorates awarded to underrepresented minorities rose from 3% to 5% between 1983 and 2003; the proportion for Asians increased from 2% to 4% in that same period. Experts predict that the greatest increase in the growth in enrollment of S&E students will be from minority students. In addition, women and minorities now constitute a robust part of the S&E workforce; women make up 24.7% of the S&E workforce while blacks comprise 6.9%, Hispanics 3.2%, and Asians 14%. That is, blacks, Hispanics, and Asians combined constitute 24.1% of the S&E workforce (National Science Board, 2006).

Although this author's studies (which will be reported in greater detail below) find evidence of a robust pool of minority students enrolled in college environmental programs, the workforce of environmental organizations is not as diverse as the rest of the S&E labor force. This is particularly true of the leadership in environmental organizations. Recognizing this, many in the environmental field are pushing for greater diversity in the field. Diversity is high on the agenda once again. Since 1998, several high-profile diversity conferences have been held. In 2005 alone, at least five major diversity conferences were held around the nation. However, as discussed by the 230 delegates attending the Minority Environmental Leadership Development Initiative's (MELDI) National Summit on Diversity in the Environmental Field, held at the University of Michigan's School of Natural Resources and Environment, quite often those advocating for greater diversity in the environmental field are still confronted with phrases like, "we can't find any 'qualified' minorities," "no minorities applied for the job," "they don't stay if we hire them," "we don't know where to find minority applicants," "we hired one but it didn't work out," "we can't afford them," "we would like to hire them, but ...," and so on. In her capacity as Program Director for MELDI, the author encountered these utterings on innumerable occasions while developing the program and organizing the 2005 National Summit on Diversity in the Environmental Field.[4]

DIVERSITY AND ENVIRONMENT: TWO COMPETING FRAMES

As the above discussion shows, minority participation in the environmental movement has been characterized by two competing frames. On the one hand, leaders of mainstream environmental organizations, believing there is a limited talent pool of potential minority workers, portray minorities as unqualified for environmental jobs, unwilling to apply for them, and desirous of salaries too high for the environmental NGOs to afford. These perceptions are bolstered by scholarly research claiming that minorities are uninterested in the environment and are lacking in knowledge of and concern for the environment. In assessing whether there are structural barriers in environmental NGOs that limit the hiring, retention, and promotion of minorities, some leaders of environmental NGOs perceive none. However, environmental justice activists and scholars articulate a competing frame. They argue that racist and exclusionary hiring practices explain the demographic characteristics of the environmental workforce. They also argue that minorities are qualified, willing and able to work in environmental NGOs and agencies. The environmental justice claims are supported by studies showing that a robust pool of qualified minority applicants exists and that minorities have salary expectations that fall within the range of the salaries paid by environmental organizations and agencies.

Furthermore, leaders of mainstream environmental NGOs perceive minorities as being too concerned with social justice and poverty-related issues. Such leaders make a distinction between social issues and environmental issues. Scholars make a similar argument when explaining what is seen as the lack of participation of minorities in environmental affairs by citing minorities' preoccupation with their basic needs and their inability to elevate their concerns to focus on the environment. In contrast, the environmental justice frame deliberately links the environment and social justice concerns and infuses such ideological thinking into the way agendas are set and articulated, and organizations are structured and operated.

A NEW LOOK AT INSTITUTIONAL DIVERSITY

Despite decades of research on the topic, we still have limited information on the status of diversity in environmental institutions. What are the demographic characteristics of contemporary environmental organizations and what kinds of diversity initiatives are underway in these organizations? Not withstanding the increased participation of minorities in environmental affairs, questions still linger about whether minorities are qualified to work in environmental organizations, their preparation for the workforce, willingness to work in environmental nonprofits, salary expectations, and the salience of diversity. To investigate these issues, the author conducted three national studies. They are: (1) a study of environmental organizations – mainstream, environmental justice, and government environmental agencies; (2) a study of white and minority students in college environmental programs; and (3) a study of the work experiences of white and minority workers in the environmental workforce. A brief discussion of some of the findings of the first two studies will follow.

The Workforce

Between 2004 and 2006, interviews were conducted with leaders of 243 environmental organizations to find out about the demographic character- istics of their leadership, staff, membership, and boards; recruitment, hiring, and retention of workers; wage scales; review and promotion practices; training and leadership development; mentoring; and diversity programming. The sample consisted of 166 mainstream environmental NGOs, 39 environ- mental justice organizations, and 38 government environmental agencies.

The study revealed that women were well represented on the staff of the organizations studied: 97.7% had females on the staff. However, ethnic minorities were still underrepresented in this sector of the workforce; overall, they comprised 17.2% of the staff of the organizations studied (see Table 1). Table 1 also shows that blacks comprise 5.7% of the sample of 20,218 staff in all three types of organizations, Hispanics 4.8%, Native Americans 4.7%, and Asians 2%. When the racial composition of the staff of each kind of organization was analyzed, mainstream environmental organizations were found to have the lowest percentage of minority staff. Minorities were found to comprise 14.6% of the staff of mainstream environmental organizations, 15.4% of government environmental agencies, and 77.8% of environmental justice organizations. Although the percentage of minorities did not differ significantly between mainstream environmental organizations and government environmental agencies, the distribution of each ethnic minority group varied substantially between the two types of organizations. A much higher percentage of Asians were found in mainstream environmental organizations than were found in government environmental agencies and environmental justice organizations. Native Americans were far more likely to be hired in government environmental agencies (most likely the Department of the Interior's Bureau of Indian Affairs) than in mainstream environmental organizations. While Native Americans comprised less than 1% of the staff of mainstream environ- mental organizations sampled, Asians constituted less than 1% of the staff of the government environmental agencies sampled.

Table 1.　Percentage of Minority Staff in Three Types of Environmental Organizations.

Types of Environmental Institutions	Total			Percentage of Minorities on Staff in Environmental Institutions							
				Black		Hispanic		Native American		Asian	
	Total Number of Staff	Total Number of Minorities	Percent Minority	Number	Percent	Number	Percent	Number	Percent	Number	Percent
All organizations combined	20,218	3,480	17.21	1,151	5.69	968	4.79	947	4.68	414	2.05
Mainstream environmental organizations	7,200	1,050	14.58	353	4.90	339	4.71	60	0.83	298	4.14
Government environmental agencies	12,334	1,898	15.38	521	4.22	407	3.30	865	7.01	105	0.85
Environmental justice organizations	684	532	77.79	277	40.5	222	32.46	22	3.22	11	1.61

Table 2. Minority Hiring in Three Years Prior to Being Interviewed for the Study.

Types of Environmental Institutions	Minority Hiring in Environmental Institutions		
	Percent Having No Minorities on Staff at Time of Interview	Hired Staff in Three Years Prior to Interview	Hired No Minority Staff in Three Years Prior to Interview
All organizations combined	28.4	92.7	28.6
Mainstream environmental organizations	34.5	93.1	33.9
Government environmental agencies	19.4	85.7	29.2
Environmental justice organizations	11.4	85.3	3.7

As Table 2 shows, 34.5% of the mainstream environmental organizations indicated that they had no minorities on staff, so did 19.4% of the government environmental agencies and 11.4% of the environmental justice organizations. Although most of the institutions hired staff in the three years leading up to being interviewed for the study, a substantial number of mainstream environmental organizations and government environmental agencies did not hire any new minority staff. Ninety-three percent of the mainstream environmental NGOs, 85.7% of the government environmental agencies, and 85.3% of the environmental justice organizations reported hiring staff in the three years prior to being interviewed for the study. Of those hiring staff, 33.9% of the mainstream environmental NGOs, 29.2% of the government environmental agencies, and 3.7% of the environmental justice organizations did not hire any minority staff in the three years leading up to the time of the interview.

An analysis was also conducted to find out the number of minorities in each type of organization. Results show that 41.1% of mainstream environmental organizations had no black staff, almost half did not have any Hispanic staff, a little more than half had no Asian staff, and 78.4% had no Native Americans on staff (see Table 3). In contrast, 15.4% of the government environmental agencies had no blacks, almost 21.7% had no Hispanics, 36% had no Asians, and 38.1% had no Native Americans on staff.

Table 3. The Relative Amounts of Each Ethnic Minority Group Found on the Staff in Each Type of Institution.

Types of Environmental Institutions	Percentage of Organizations Having the Specified Number of Ethnic Minorities on Staff											
	Black			Hispanic			Native American			Asian		
	None	10 or Fewer	24 or Fewer	None	10 or Fewer	24 or Fewer	None	10 or Fewer	24 or Fewer	None	10 or Fewer	24 or Fewer
All organizations combined	34.8	86	93.3	45.8	90.3	95.5	71.5	96.7	96.7	53.5	94.2	98.7
Mainstream environmental organizations	41.1	90.7	97.2	49.5	92.2	99	78.4	100	100	53.8	94.2	99
Government environmental agencies	15.4	73.1	84.6	21.7	82.6	87	38.1	76.2	76.2	36	88	96
Environmental justice organizations	29	87.1	93.5	51.7	89.7	89.7	71.4	100	100	69.2	100	100

Table 4. Demographic Characteristics of Organizations in the Study.

Leadership Positions	Percent Having Position	Percent Gender and Race for Each Position					
		Gender			Race		
		Male	Female	Both	White	Minority	Both
Chief executive officer	97	62.8	30.5	6.7	86	12.2	1.8
Chair of the board	73.3	67.6	30.4	2	83.7	15	1.4
Secretary	73.4	32.7	65.4	1.9	72.7	22.7	4.5
Treasurer	59.3	54.1	45.1	0.8	74.8	23.6	1.6
Accountant	65.4	27.7	62.8	9.5	70.1	18.2	11.7
Program/activities director	81.7	28.3	34.1	37.6	71.8	11.5	16.7
Youth director	25.3	34	53.2	12.8	66	23.4	10.6
Community organizer	54.4	17.9	57.5	24.5	66	24.5	9.4
Publications editor	62.4	34.4	56.4	9	85.8	8.3	5.8
Public relations manager	49.5	38.3	54.3	7.4	81.7	14	4.3
Spokesperson	26.9	53.2	42.6	4.3	71.7	26.1	2.2
Lobbyist	24.9	37	23.9	39.1	75	11.4	13.6
Lawyer, legal counsel	38.6	47.1	27.1	25.7	80	10	10
Diversity manager	21.2	31.6	60.5	7.9	30.8	61.5	7.7

The Leadership

Whites and males still dominate the top leadership positions in the organizations. Table 4 shows the percent of organizations that have each leadership position and the racial and gender composition of each position. Table 5 compares the differences and similarities of the leadership structure of the three types of institutions. Overall, 63% of the chief executive officers (CEOs – this also includes president, executive director, etc.) were male and 86% were white. In mainstream environmental organizations, 64% of the CEOs were male and 96% were white. Women and minorities had the highest chances of being CEOs in environmental justice organizations; even though a majority of the CEOs were male; 53% were male and minority.

Generally speaking, about 68% of the chairs of the board were male and roughly 84% were white. As was the case for CEOs, a higher percent of the chairs of the board were male and white in mainstream environmental organizations than in environmental justice organizations. Only 59% of

Table 5. Leadership Structure of Various Environmental Organizations.

Leadership Positions	Mainstream Environmental Organizations							Government Environmental Agencies							Environmental Justice Organizations						
	Percent Having Position	Percent Gender and Race for Each Position						Percent Having Position	Percent Gender and Race for Each Position						Percent Having Position	Percent Gender and Race for Each Position					
		Gender			Race				Gender			Race				Gender			Race		
		Male	Female	Both	White	Minority	Both		Male	Female	Both	White	Minority	Both		Male	Female	Both	White	Minority	Both
Chief executive officer	97.5	64.3	30.5	5.2	95.5	3.2	1.3	92.1	66.7	24.2	9.1	87.5	9.4	3.1	100	52.8	36.1	11.1	44.4	52.8	2.8
Chair of the board	83.6	69.2	29.1	1.7	89.7	8.5	1.7	61.1	66.7	33.3		100			34.6	59.1	36.4	4.5	42.9	57.1	
Secretary	76.7	35.1	62.3	2.6	82.1	13.4	4.5	66.7	11.1	88.9		66.7	22.2	11.1	64.3	37.5	62.5		33.3	66.7	
Treasurer	67.4	56.8	42.1	1.1	83.2	14.7	2.1	52.8	28.6	71.4		62.5	37.5		25.9	50	50		40	60	
Accountant	66.2	25.5	62.8	11.7	77.4	9.7	12.9	56.8	23.8	66.7	9.5	61.9	23.8	14.3	72.4	40.9	59.1		47.8	47.8	4.3
Program director	85.7	27.2	34.4	38.4	80	6.4	12.8	75	42.9	9.5	47.6	61.9	4.8	33.3	70	22.2	51.9	25.9	39.3	39.3	21.4
Youth director	27	32.4	58.8	8.8	76.5	14.7	8.8	22.2	60	20	20	50	25	25	20.8	25	50	25	33.3	55.6	11.1
Community organizer	54.9	13.7	63	23.3	82.2	12.3	5.5	57.1	41.7	41.7	16.7	53.8	23.1	23.1	48	19	47.6	33.3	15	70	15
Publications editor	72.6	29.6	60.2	10.2	92.9	2	5.1	29.4	61.5	38.5		66.7	25	8.3	52	45.5	45.5	9.1	40	50	10
Public relations manager	52	33.3	57.6	9.1	90.9	3	6.1	29.4	47.1	47.1	5.9	76.5	23.5		63	54.5	45.5		30	70	
Spokesperson	25.4	41.4	51.7	6.9	89.7	6.9	3.4	23.5	88.9	11.1		77.8	22.2		39.1	55.6	44.4		100		
Lobbyist	30.2	32.4	24.3	43.2	78.4	8.1	13.5	11.8	50	25	25	100			16	60	20	20	25	50	25
Lawyer, legal counsel	40.8	45.1	23.5	31.4	82.4	5.9	11.8	32.4	50	50		100			36	54.5	27.3	18.2	54.5	36.4	9.1
Diversity manager	16.8	20	75	5	40	55	5	5.7	43.8	43.8	12.5	17.6	70.6	11.8	64	50	50		50	50	

them were male in environmental justice organizations and 57% were minority. The positions of secretary, accountant, youth director, community organizer, and diversity manager were dominated by women. Whites also dominated these positions in mainstream and government agencies. Although women dominated program directors positions in mainstream environmental organizations and environmental justice organizations, males held a high percentage of these positions in government agencies.

The diversity manager position is of interest for a number of reasons. In the 1992 ECO survey, leaders of environmental organizations indicated that diversity was a salient issue in their organization and that they wanted to diversify their organizations. Nonetheless, more than a decade later, only 17% of the mainstream environmental organizations in this study indicated that they had a diversity manager. In contrast, 44% of the government environmental agencies had a diversity manager. Women dominated this position in all three types of organizations. Furthermore, this is the only leadership post that minorities held in significant amounts in mainstream organizations. Seventy-five percent of the diversity managers in mainstream environmental organizations were women and 55% of them were minority. In government agencies, the position seem to be evenly split between males and females, but 70% of the diversity managers were minorities. Very few environmental justice organizations had diversity managers. This is probably partly due to size – many of the organizations were small. Moreover, many of the environmental justice organization were founded on cross-race, cross-class partnerships, alliances, memberships, and principles. Consequently, diversity was a core organizing principle for them.

Tables 4 and 5 also show that with the exception of the diversity manager's position, low percentage of minorities hold top leadership positions in environmental institutions. In 13 of the 14 leadership positions studied in mainstream environmental organizations, more than 75% of the people holding those positions were white. Minorities seem to have a higher chance of occupying leadership positions in government environmental agencies. Although more than three quarters of the people holding the CEO, chair of the board, public relations manager, spokesperson, lobbyist, and legal counsel positions in government environmental agencies were white – more than 25% of the staff holding the remaining eight positions were minorities. The reverse is true for environmental justice organizations, where minorities held at least 50% of the positions for 10 of the 14 types of posts analyzed.

The more prominent and significant the leadership position in the organizations, the more unlikely it seemed that minorities would hold those posts. For instance, 12% of the CEOs from the three types of organizations were minorities (another 2% of the organizations had both white and minority CEOs), but less than 5% of the mainstream environmental organizations sampled have minority CEOs. Nine percent of the government environmental agencies have a minority CEO and another 3% had both white and minority CEOs. Similarly with the chair of the board position, less than 10% of the chairs of the boards in mainstream environmental organizations were minority. None of the chairs of the boards in government environmental agencies were minority.

What frame best explains the low percentage of minorities employed in mainstream environmental organizations as well as government environmental agencies? How does one explain the low levels of hiring of minorities in these institutions? Are the figures detailed above a reflection of lack of minorities in the pipeline? Is it unwillingness to work in mainstream and government organizations or minorities making salary demands that are too high? Or, is it a function of recruitment practices? These questions will be explored further by analyzing data collected from the national survey of students in environmental programs.

MINORITY STUDENTS AND THE WILLINGNESS TO WORK IN ENVIRONMENTAL ORGANIZATIONS

A study of students enrolled in four-year college S&E programs was conducted between 2003 and 2005. Surveys were administered to students in S&E programs that were environmental in nature. The study focused on students in five life sciences fields (biological sciences, forestry, natural resource management, agricultural sciences, and environmental sciences), a physical science field (geosciences – geology, earth science and atmospheric science), an engineering field (environmental engineering), and two social science groups (geography and students pursuing degrees like environmental sociology and political ecology). The study examined students' willingness to work in five types of institutions upon graduation: academia, government agencies, corporations (for-profit organizations), environmental organizations, and other nonprofits.

Of the 1,224 students identifying their racial backgrounds, five were mixed race or Arabs. Because the number of mixed race and Arab respondents were so few, these respondents were excluded from the analysis. Of the remaining 1,219 respondents, 348, or 28.5%, were members of minority groups, and 871, or 71.5%, were white. Further analysis of the sample shows that 10% (122) of the respondents were Asian, 9.1% (111) were Latino,

7.9% (96) were black, and 1.6% (19) were Native American. The sample contained 708 females and 506 males; it also had 501 doctoral students, 515 master's, and 212 undergraduates. Forty-six percent of the undergraduates were minorities, so were 21.6% of the master's, and 28.2% of the doctoral students. Sixty-five percent of the undergraduate respondents were female, so were 62.3% of the master's, and 51.7% of the doctoral students.

Respondents were asked to rate their likelihood of working in a particular type of institution upon graduation on a six-point Likert scale, with 0–1 being classified as unlikely and 2–5 being classified as likely. Analysis of the sample mean scores showed that working in a government environmental agency and teaching in academia had the highest means (3.10 and 3.09, respectively). Environmental justice organizations had the lowest mean score of 1.79. In terms of percentages, 84.7% of the sample indicated they would be willing to work in a government environmental agency upon graduation (see Table 6). Academic institutions were next; 79.1% of the respondents indicated they would be willing to teach and 78.7% indicated a willingness to work as research scientists in such institutions. Although the percentages were lower, almost three quarters of the respondents indicated a willingness to work in mainstream environmental organizations. The institutions that were the least attractive to respondents were environmental justice organizations (52.1%) and corporations without an environmental division (49.2%).[5]

Fig. 1 and Table 6 show the percentages of whites and minorities indicating a willingness to work in particular types of institutions. Although the analysis of whites and minorities show almost identical percentages of respondents willing to work in mainstream environmental organizations or as researchers in academia, more detailed analysis shows significant variation among different racial groups in the sample. There was noticeable difference between blacks and other groups when the relationship between race and the willingness to teach in academic institutions was examined – 82.5% of Asians, 79.9% of whites, and 78.2% of Hispanics were willing to work as teachers in academic institutions. In comparison, 69.8% of blacks expressed such willingness. The variation was even more dramatic when willingness to research in academia was examined. Asians (88.2%) were far more likely than any other group to indicate willingness to research in academic institutions. Native Americans (89.5%) were most likely to say they would work for mainstream environmental organizations, and Asians (63%) were most likely to indicate willingness to work for environmental justice organizations. Blacks (63.2%) were far less likely than others to say they were willing to work for mainstream environmental organizations. Whites (49.7%) were the least likely to indicate a willingness to work for environmental justice organizations.

Table 6. Race, Gender and the Willingness to Work in Different Types of Institutions.

Factors	Mean and Percent of Respondents Willing to Win Particular Types of Institutions								
	Academia Teaching	Academia-research Scientist	Mainstream Environmental Organization	Environmental Justice Organization	Other Nonprofits	Government Environmental Agency	Other Government Agency	Corporate Environmental Organization	Other Corporate Entities
Total mean	3.09	3.01	2.52	1.79	2.21	3.10	2.72	2.25	1.83
Total sample percent	79.1	78.7	73.6	52.2	67	84.8	76.3	64.3	49.3
Race									
White	79.9	78.3	73.6	49.7	64.8	85.5	74.9	62	42.8
All minorities combined	77.1	79.7	73.5	58.3	72.3	82.9	79.9	70.4	65.6
Black	69.8	68.8	63.2	53.1	72.9	85.4	92.6	78.9	83.9
Native American	73.7	73.7	89.5	57.9	78.9	78.9	68.4	63.2	63.2
Hispanic/Latino	78.2	80.9	75.5	57.7	69.1	84.5	74.3	58.6	47.3
Asian	82.5	88.2	77.3	63	73.7	80	76.7	75.8	68.6
Gender									
Male	80.4	82.1	69.9	48	61.6	84.5	81.6	68.6	55.2
Female	78.5	76.1	76.3	55.1	71	85	72.6	61.2	44.9

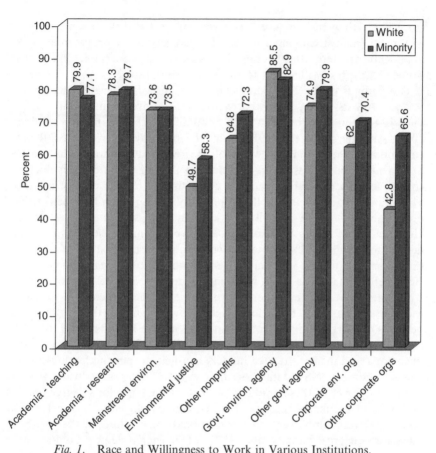

Fig. 1. Race and Willingness to Work in Various Institutions.

Native Americans (78.9%) were most likely and whites (64.8%) least likely to indicate a willingness to work in non-environmental nonprofits. While 78.9% of Native Americans and 80% of Asians said they were willing to work in government environmental agencies, approximately 85% of the other respondents said the same. Blacks (92.6%) were far more likely than others to say they were willing to work in non-environmental government agencies. Blacks were also most likely to say they were willing to work in corporations upon graduation.

Women were less likely than men to say they would work in academia as teachers or researchers. Gender differences were also apparent in questions exploring willingness to work in nonprofits and corporations. Women were more likely than men to say they were willing to work in three types of nonprofits (mainstream environmental organizations, environmental justice organizations, and other nonprofits) while men were more likely than women to express willingness to work in corporations. The findings of the 2004 American Association for the Advancement of Science (AAAS) salary survey corroborate the above findings; only 15% of the 6,124 scientists in the AAAS survey studied worked in industry (Holden, 2004). Another study, the 1997 Survey of Doctorate Recipients (SDR) also corroborates these findings. Most of the recent doctorate holders in the SDR who wanted to work in academia at the beginning of their doctoral studies actually worked in academia at the time of the survey. For instance, 61% wanted to work in a college/university setting when they began their doctorate; 48% were actually working in that setting when surveyed. However, more of the SDR respondents worked in industry than initially desired to – while 24.3% wanted to work in business/industry at the start of their doctoral studies, 32% actually ended up working in business/industry. Four percent had a desire to work in government and 9.2% were actually working in government at the time the survey was conducted. In the SDR, the geoscientists and biological scientists were among the groups most desirous of working in academia (National Science Foundation, 2001). The same was true for this study. However, other NSF data on recent S&E bachelor's and master's degree holders can help to put the findings of this study in context. NSF data show that in 2003, business/industry was the largest employer of these graduates – 68% of the bachelor's degree holders and 58% of the master's degree holders were employed in this sector. The second largest employer was educational institutions; 21% of S&E bachelor's degree holders and 28% of master's degree holders were employed in this sector. The government sector employed 11% and 14% of S&E bachelor's and master's degree holders, respectively (National Science Foundation, 2005, 2006). The findings of the student survey corroborate the NSF employment data in that 68.1% of undergraduates and 69.7% of masters' students in this sample said they were willing to work in corporate environmental entities and more than half said they would work in other corporations.

Data from the Bureau of Labor Statistics (BLS) that provide information on environmental jobs were also examined. BLS data show that more than a half of the biological scientists, conservation scientists, foresters, forest technicians, social scientists, environmental scientists, and hydrologists work for a government entity. Sixty-one percent of chemical science technicians worked in corporations, so did 29% of environmental engineers,

27% of biological technicians, 18% of agricultural technicians, about 45% of geoscientists, 23% of geosciences technicians, and 29% of environmental scientists (Bureau of Labor Statistics, 2005). Hence, BLS data indicate that the preference of respondents to work in government is realistic.

RECRUITMENT

Minority students may be willing to work in environmental organizations, but are they being recruited? Recruitment is an integral part of diversity efforts, but to what extent are organizations identifying and recruiting students in the pipeline? Only 32% of the students in the sample report being recruited for a job by environmental organizations in the last five years. Although 53% of Native American students report being recruited for jobs, only 35% of Asians, 33% of blacks, 32% of Hispanics, and 31% of whites report being recruited for jobs in environmental organizations. It seems as though environmental organizations have to do a better job of identifying and recruiting students in the pipeline for open positions in their organizations.

How are the environmental organizations recruiting new workers? The 243 environmental organizations discussed above also reported on their recruitment practices. The most mechanisms used are: recruiting through websites (33.9%), newspapers (28.1), from within environmental networks (20.1%), in the local community (19.2%), by word of mouth (16.5%), and through local and state universities (14.3%). Only 9.8% of the organizations conducted national searches. Mainstream environmental organizations recruited most heavily through websites (41.3%), newspapers (37.4%), from within environmental networks (22.6%), and word of mouth (20%). On the other hand, government environmental agencies relied more heavily on recruiting from local and state universities (32.4%) and from the local community (14.7%); only11.8% recruited through websites or from within environmental networks.

THE SALARY EXPECTATIONS OF MINORITIES

Students were asked to state the minimum salary they would accept upon graduation. The mean minimum salary was calculated. However, to facilitate detailed analyses, responses were grouped into five categories – $0–$29,000, $30,000–$39,000, $40,000–$49,000, $50,000–$59,000, and

$60,000 and over. Generally speaking, respondents had modest salary expectations that were comparable to what was being paid to new S&E graduates. The mean minimum salary expected by the sample was $39,371. Fifty-one percent of the respondents said they would accept less than $40,000 as their minimum salary upon graduation. Only 11% of the respondents said they wanted a minimum salary of $60,000 or more (see Table 7).

When race was analyzed, Asian students generally expected higher wages than other students. Asians had the highest mean minimum acceptable salary of $44,954; the white students' mean of $38,294 was the lowest. While 55.5% of Native American, 54.6% of white, 49.5% of Hispanic, and 41% of black students would accept minimum salaries that were less than $40,000, only 30.6% of Asian students said they would do likewise. In a similar vein, Asian students were more likely to indicate they would accept a minimum salary that was $50,000 or more; 38.8% of Asian students felt this way. In contrast, less than 30% of the other students expected a minimum salary in this range.

This author's organizational study found that the mean salaries being paid to new employees with a bachelor's degree ranged from $17,000 to $52,500; the mean was $30,374. The mean wage being paid to new employees with a master's degree was $37,401 and $51,889 to those with a Ph.D (see Table 8).

Table 7.   Students' Salary Expectations.

Factors	Percent Indicating Minimum Acceptable Salary					Mean Minimum Acceptable Salary
	$0–$29,000	$30,000–$39,000	$40,000–$49,000	$50,000–$59,000	$60,000 and over	
Total	15.5	35.5	26.8	11.2	11	$39,371
Race						
White	16.5	38.1	26.2	10.1	9	$38,294
All minorities combined	13.4	27.7	29.3	13.1	16.5	$42,250
Black	10.5	30.5	29.5	12.6	16.8	$43,095
Native American	22.2	33.3	22.2	11.1	11.1	$40,361
Hispanic/Latino	18.1	31.4	27.6	13.3	9.5	$39,029
Asian	9.3	21.3	30.6	15.7	23.1	$44,954
Gender						
Male	14.1	29.8	30.4	12	13.7	$41,275
Female	16.7	38.9	24.4	10.7	9.2	$38,115

The mean wages being paid to new employees varied somewhat in the different types of organizations. While government environmental agencies paid the lowest mean wages to those with a bachelor's degree, the mean wages being paid to new employees with a masters and doctorate was substantially higher in government environmental agencies than in mainstream environmental organizations or environmental justice organizations. While the mean wages being paid by environmental justice organizations to new employees with a bachelors degree is similar to that being paid by government environmental agencies and mainstream environmental organizations, the mean wages being paid by environmental justice organizations to new employees with a masters or doctorate was significantly lower than it was in the other two types of organizations. The mean minimum salary that minority students in this study indicated would be $36,489 for undergraduates, $42,380 for those in master's programs and $45,876 for doctoral students. The mean minimums for white students were $32,270 for undergraduates, $35,963 for master's students and $42,651 for Ph.D. students.

The data indicate that there is a robust pool of minority students whose salary expectations fall within the range of that being paid by environmental institutions. In other words, 92% of the government environmental agencies, 90% of the mainstream environmental organizations and 88% of the environmental justice organizations paid new workers with a bachelor's degree less than $40,000 to start. In comparison, 64.3% of minority and 77.5% of white undergraduates would accept a minimum salary of less than $40,000. While 69% of the environmental justice organizations, 61% of the government environmental organizations and 60% of the mainstream environmental organizations offered new workers with a masters' degree less than $40,000 to start, 39.4% of minority and 64.7% of white master's degree students said they would accept a minimum salary in this range. All the government environmental agencies and 77% of the mainstream environmental organizations were paying new staff with doctorates between $40,000 and $59,000. This matched the salary expectations of most minority and white doctoral students. The environmental justice wage scale for doctorates was lower than white and minority students in the sample indicated would be acceptable minimum wages.

How do the salary expectations of respondents in the sample compare to others in the general population and to salaries being paid to new college graduates? The National Association of Colleges and Employers reports that among recent graduates with bachelor's degrees, engineers could expect

Table 8. Mean Salaries being Paid by Environmental Organizations to New Employees with Different Types of Degrees.

Types of Organizations	Undergraduates			Masters			Doctorates		
	Lowest Mean Reported	Highest Mean Reported	Mean	Lowest Mean Reported	Highest Mean Reported	Mean	Lowest Mean Reported	Highest Mean Reported	Mean
All types of organizations combined	$17,000	$52,500	$30,374	$20,000	$65,000	$37,401	$20,000	$100,000	$51,889
Government environmental agencies	$20,000	$40,000	$29,840	$25,000	$60,000	$39,306	$40,000	$100,000	$58,083
Mainstream environmental organizations	$17,000	$52,500	$30,559	$24,000	$65,000	$37,719	$27,500	$85,000	$51,898
Environmental justice organizations	$20,000	$42,500	$30,150	$20,000	$45,000	$32,585	$20,000	$50,000	$32,125

to starting salaries over $50,000, those in the geosciences about $40,000, agriculture around $36,000, the biological sciences about $30,000, and conservation/forestry about $27,000 (NACE, 2006). It seems that most of the undergraduates in the author's student study being reported have salary expectations that are within the range of what's being paid by employers – i.e., 64.3% of the undergraduates in the sample said they would accept a minimum salary of less than $40,000 upon graduation.

A 2004 *Wall Street Journal* poll of 1,000 undergraduates found that 45% of the respondents expected to earn $30,000 or more upon graduation; 21% of those surveyed expected starting salaries of $40,000 or more (Kim, 2004). In comparison to the students in the *Wall Street Journal* poll, 66% of the undergraduates in the study being discussed here said the minimum salary they would accept would have to be $30,000 or more. In addition, 29.2% said the minimum salary they would accept would be $40,000 or more. Students in this study had somewhat higher salary expectations than students in the *Wall Street Journal* sample. This could be due to the fact that this study has a sample of S&E undergrads while the *Wall Street Journal* poll was conducted amongst undergraduates from a broader range of academic fields.

THE SIGNIFICANCE OF DIVERSITY AND EQUITY IN ENVIRONMENTAL ORGANIZATIONS

Students were asked to rate 20 indicators of institutional diversity and equity on a six-point Likert scale, with zero indicating the factor was not at all important and 1–5 indicating its significance from marginally important (1) to extremely important (5). They were asked to say how important they thought each factor would be in their decision to work for an organization. For the first round of analysis responses were grouped into four categories: zero = not at all important, 1–2 = marginally important, 3 = important, and 4–5 = very or extremely important.

Ninety percent or more of the students thought all the factors would be of some importance in their decision to work in an organization.[6] Almost all thought that variables such as fairness in promotions, pay equity for equal education and work experience, the ability to collaborate with colleagues, and being allowed to take on leadership roles would have some importance in their decisions. While there was little variance in the sample at this level of

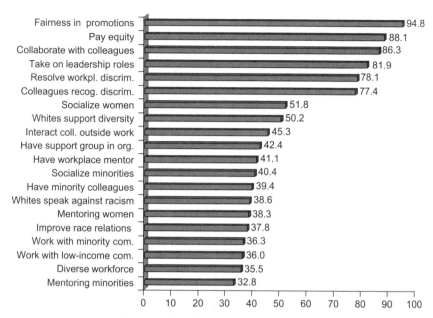

Fig. 2. Percentage of Students Considering each Factor Very/Extremely Important.

analysis, as Fig. 2 shows, there were dramatic differences in the percentage of students who thought each factor was very/extremely important.

Fig. 2 shows that more than three quarters of the respondents thought that fairness in promotions (94.8%), pay equity (88.1%), ability to collaborate with colleagues (86.3%), and take on leadership roles (81.9%), the organization's efforts to resolve workplace discrimination (78.1%), and having colleagues who recognize discrimination (77.4%) would be very/ extremely important. It should be noted that these are all institutional characteristics from which everyone in a given organization would benefit if they were in place. A specific subgroup of workers was not mentioned in any of these variables. Variables that mentioned subgroups of workers were viewed as very/extremely important by much smaller percentages of respondents. Moreover, the smaller the subgroup mentioned, the smaller the percentage of respondents identifying it as very/extremely important. Hence, variables mentioning females, such as the socializing of women (51.8%) were considered very/extremely important by more respondents than variables like socializing minorities (40.4%) or mentoring minorities (32.8%).

Who Considered each Diversity and Equity Factor to be Very or Extremely Important?

More in-depth analysis was performed to identify which respondents gave a score of 4 (very important) or 5 (extremely important) to each factor. A dichotomous dependent variable was created for each of the 20 factors by collapsing the categories on the Likert scale, giving a score of zero if respondents did not consider the factor very/extremely important and one if they did. Racial differences in how significant respondents thought each factor would be in their decision were quite pronounced when the question of who considered the dependent variables to be very or extremely important was considered (see Fig. 3). The racial differences were even more pronounced if small subgroups of workers were mentioned. White respondents seemed to make the sharpest distinctions between variables that mentioned subgroups of workers and those that did not. Thus, whites were

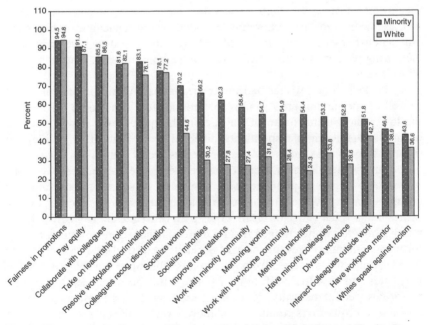

Fig. 3. Race and the Percent of Students Considering Diversity and Equity Issues Very or Extremely Important in the Workplace.

far more likely to say that factors like fairness in promotions and pay equity were very/extremely important than they were to say that socializing and mentoring minorities were very/extremely important. On the other hand, minorities seemed less inclined to distinguish between variables based on this criterion. Consequently, higher percentages of minorities than whites perceived the factors as being very or extremely important regardless of whether it mentioned specific groups.

Comparing Means

A means comparison shows striking variations in the sample according to race and gender (see Table 9). First, the factors with the highest means (on a six-point scale) were: fairness in promotions (4.65), pay equity (4.45), ability to collaborate with colleagues (4.33), being able to take on leadership roles (4.20), the organization resolving workplace discrimination (4.12), and having colleagues recognize discrimination (4.10). The factors with the lowest means (less than 3.0 on the scale) were: diverse workforce, working with minority communities, having white colleagues who speak out against racism, organization's efforts to improve race relations, mentoring women, working with low-income communities, and mentoring minorities.

As Table 9 shows, whites had lower means than all minorities combined as well as specific minority groups for most factors considered. Moreover, the lower the general mean of the factor, the more likely it was that the mean score for white respondents diverged from that of minorities. In these instances, the mean scores for whites were noticeably lower than that of minorities. Racial differences were even more pronounced when the responses of white respondents were compared to specific racial minority groups – blacks, Native Americans, Hispanics, and Asians. In other words, whites had a wider range of mean scores than other racial groups. While the mean scores for whites ranged from 4.65 for fairness in promotions to 2.49 for mentoring minorities the means for blacks ranged from 4.66 to 3.11, for Native Americans it ranged from 4.63 to 3.16, the range was 4.80 to 3.03 for Hispanics and 4.54 to 2.98 for Asians. The differences between whites and minorities were statistically significant for all but four dependent variables (fairness in promotions, ability to collaborate with colleagues, being allowed to take on leadership roles, and having colleagues who recognize discrimination).

ARE MINORITIES QUALIFIED TO WORK IN ENVIRONMENTAL ORGANIZATIONS?

The author tried to assess qualifications for the environmental workforce by looking at indicators of workforce preparation. Six indicators of workforce preparation were examined. White and minority students were compared on each indicator. The findings show that preparation of each group is relatively similar (see Table 10). More than 80% of the students had field experiences and attended professional conferences. The activity that students were least likely to participate in was organizing conferences. The differences between minority and white student on indicators of workforce preparation were so close that they were statistically insignificant.

Mentoring and Career Outcomes

This section of the essay will discuss mentoring and its relationship to career outcomes in more detail. This is important because mentoring enhances the retention of students, academic success and career development and advancement. Studies have found that mentoring is positively related to career success. That is, people who have been mentored tend to be promoted more frequently, earn higher incomes, and were more satisfied with their income and benefits than those who had few or no mentoring relationships. Mentoring works because it provides role models while helping to socialize and integrate protégés into norms and practices of the workplace. It fosters networks that help protégés to bridge the structural holes in their work environment. Mentoring increases communication in organizations, the early recognition of potential leaders, productivity, and decreases turnover (Eshner & Murphy, 1997; Dreher & Ash, 1990; Fagenson, 1989; Wilson & Elman, 1990; Wallace, 2001).

Most of the research on mentoring has focused on the experiences of white males. However, in recent years, researchers have begun examining the mentoring experiences of women and minorities (Eshner & Murphy, 1997; Maniero, 1994; Wallace, 2001; Sosik & Godshalk, 2000; Thomas & Alderfer, 1989; Dreher & Chargois, 1998). Researchers found that white female executives and black managers had multiple mentors instead of the traditional mentor-protégé dyad. A comparative study of black and white managers found 62% of blacks had both white and black male mentors (Maniero, 1994; Thomas & Alderfer, 1989; Eshner and Murphy, 1997).

Table 9.   Means Comparison: Students' Ranking of importance of Diversity and Equity Factors.

Independent Variables	Students' Mean Ranking of Each Diversity and Equity Issue								
	Fairness in Promotions, Advancement	Pay Equity for Equal Education, Experience	Ability to Collaborate with Colleagues	Being Allowed to Take on Leadership Roles	Organization Resolving Workplace Discrimination	Having Colleagues Recognize Discrimination	Organization's Efforts to Socialize Women	Interacting with Colleagues Outside Work	Having White Colleagues Support Diversity
Total sample mean	4.65	4.45	4.33	4.20	4.12	4.10	3.31	3.26	3.26
Race									
White	4.65	4.42	4.33	4.19	4.08	4.10	3.10	3.21	3.11
All minorities combined	4.65	4.52	4.35	4.23	4.24	4.11	3.84	3.38	3.62
Black	4.66	4.46	4.22	4.29	4.29	4.27	4.15	3.17	4
Native American	4.47	4.32	4.26	4.63	4.37	4.42	3.63	3.50	3.74
Hispanic/Latino	4.80	4.65	4.54	4.29	4.30	4.21	3.85	3.38	3.46
Asian	4.54	4.47	4.28	4.07	4.11	3.84	3.61	3.53	3.43
Gender									
Male	4.58	4.24	4.35	4.18	3.99	3.97	2.92	3.26	2.98
Female	4.70	4.61	4.33	4.23	4.22	4.19	3.60	3.25	3.45

Students' Mean Ranking of Each Diversity and Equity Issue

Having a Mentor in the Workplace	Organization's Efforts to Socialize Minorities	Having Minority Colleagues	Having a Support Group at Work	Diverse Workforce	Working with Minority Communities	Having White Colleagues Speak out Against Racism	Organization's Efforts to Improve Race Relations	Mentoring Women	Working with Low-income Communities	Mentoring Minorities
3.09	3.04	3.02	3	2.95	2.94	2.93	2.93	2.91	2.85	2.77
3.02	2.76	2.86	2.78	2.72	2.68	2.86	2.63	2.69	2.63	2.49
3.28	3.74	3.40	3.55	3.50	3.59	3.11	3.66	3.45	3.41	3.47
3.62	4.10	3.76	3.94	3.69	3.96	3.27	4.05	3.80	3.74	3.86
3.16	3.58	3.37	3.21	3.58	3.26	3.21	3.21	3.47	3.21	3.47
3.22	3.71	3.55	3.36	3.58	3.67	3.03	3.63	3.59	3.57	3.65
3.09	3.49	2.98	3.46	3.27	3.27	3.03	3.44	3.02	3.04	2.98
2.93	2.85	2.85	2.72	2.78	2.72	2.65	2.75	2.58	2.65	2.62
3.22	3.18	3.13	3.20	3.07	3.11	3.13	3.06	3.14	3.00	2.88

Table 10. Race and Indicators of Workforce Preparation.

Workforce Preparation	Total Sample		Whites		Minorities	
	Percent Yes	Mean	Percent Yes	Mean	Percent Yes	Mean
Had internships	56.3	1.15	56.2	1.10	56.7	1.27
Had field experiences	90.1	9.90	92	11.18	85.6	6.90
Have mentor	71.6	0.72	70.8	0.71	73.5	0.74
Belong to professional association	70.7	0.71	71.1	0.71	69.9	0.70
Attend professional conference – 5 years	82.4	0.83	82.5	0.82	82	0.82
Present at professional conference – 5 years	64.6	0.65	65.7	0.66	61.5	0.62
Organized conference/ workshop – 5 years	29.4	0.29	26.1	0.26	37.6	0.38

Researchers have also begun examining whether the race and gender of the mentor matters in terms of economic outcomes. The results are unclear. Some studies have found that female protégés who have male mentors get more instrumental support, advance further in their careers and report higher incomes than those with female mentors (Wallace, 2001; Dreher & Cox, 1996; Maniero, 1994; Sosik & Godshalk, 2000; Ibarra, 1992). Dreher and Chargois' (1998) work support this finding; they found black graduates who had white male mentors reported higher wages than blacks who did not have white mentors. Other studies in this genre have found that same-gender mentoring dyads provided more psychosocial support for protégés (Sosik & Godshalk, 2000; Burke, McKeen, & McKenna, 1990; Allen, Burke, & McKeen, 1997). However, some researchers have found no relationship with the gender of the mentor and protégé outcome or have failed to confirm the positive effects of female mentoring on protégé outcomes (Eshner and Murphy, 1997; Dreher & Ash, 1990; Fagenson, 1989; Ragins & McFarlin, 1990). This chapter argues that mentoring helps to prepare students for professions in the environmental field and groom them for leadership roles.[7]

The student study found that black students were the most likely and Asian students the least likely to say they had a mentor. Seventy-nine percent of black students had mentors. In comparison, 75% of Hispanics, 74% of Native Americans, 71% of whites and 69% of Asian students indicated they had a mentor. Some students had multiple mentors. While 21% of the students had only one mentor, half had multiple mentors. Black students were more likely than other students to have multiple mentors. While 61% of black students had multiple mentors, 53% of Native Americans, 52% of Asians, 51% of Hispanics and 48% of white students had multiple mentors. Studies show that professional blacks also rely on multiple mentors as well as black mentors (Eshner & Murphy, 1997; Thomas & Alderfer, 1989). It appears that black students have adopted this strategy early in their careers. These findings have larger implications. Since students are currently being mentored, many will expect this when they enter the workforce.

Most students (85%) had a mentor who was in the same discipline as they were. Interestingly enough, although black students were most likely to have one or more mentors, they were far less likely to have a mentor in their discipline than other students. Ninety-one percent of Asian students as well as 87% of white and 86% Native American students had mentors in their discipline. Moreover, 84% of Hispanics had mentors in their discipline. In contrast, only 71% of black students said their mentor was in their discipline. This might help to account for black students' heavier reliance on multiple mentors. Some may be using the multiple mentors to compensate for the guidance they miss out on by not having mentors in their discipline.

There was a positive relationship between mentoring and professional and leadership development. The study found that students who had mentors were more likely to belong to professional associations. Seventy-four percent of the respondents who had a mentor belonged to a professional association. In comparison, 65% of the students who did not have a mentor did not belong to any professional associations. The gap is even wider if the mentor is in a student's discipline. The results are similar if the mentors are in the same institution as the student. Students who had a mentor and whose mentor was in the same discipline as the student were also more likely to attend professional conferences than those who did not have a mentor or whose mentor was in another discipline. Students with mentors were also more likely to make presentations at professional conferences than other students. Students with mentors were more likely to report having internships and field experiences.

Mentoring was also positively related to minority students' professional and leadership development. This was particularly true for black, Hispanic and Asian students – when these students had mentors they are more likely to belong to professional associations than others of their racial group who did not have mentors. Those having multiple mentors were most likely to belong to professional associations. In fact, black and Hispanic students with mentors were more likely to belong to professional associations than white students. The results were similar for attendance and presentations at professional conferences. These data indicate that environmental organizations seeking to hire new employees cannot make a valid argument that minority students in this sample were so lacking in preparation for the environmental workforce that they were not qualified.

DISCUSSION AND CONCLUDING REMARKS

This chapter traced the evolution of the research on diversity in the environmental field for the last four-and-a-half decades. It addressed some of the lingering questions about minorities and the environment. The essay also discussed new data that might help us to grapple to with questions or concerns. The chapter focused on six major questions: What is the status of diversity in environmental institutions? Do minorities want to work in environmental organizations? Are minorities being recruited? Are the salary expectations of minorities too high for environmental organizations to meet? How salient is diversity in the decision to work for a particular organization? Are minorities qualified for jobs in environmental organizations?

The essay found that though there has been some progress on diversity, there is still much work needed to be done to even get environmental organizations to the level of diversity found in the remainder of the S&E workforce. With the exception of the diversity manager's position, minorities were still very underrepresented in the general staff and senior positions of most organizations. Clear evidence is presented above to demonstrate that minorities are desirous of jobs in environmental organizations and that their salary expectations are within the range of what is currently being paid.

Several factors play a role in the low levels of minority employment in mainstream environmental organizations and government environmental agencies. For decades, mainstream environmental leaders and activists have framed minorities and their relationship to the environment in negative terms. In many instances, the negative framing amounted to stereotyping that portrayed minorities as uninterested in the environment, unwilling to work for environmental organizations, desirous of salaries that are too high, and lacking in qualifications for the jobs being offered in said organizations. Although proponents of these arguments put forward little or no evidence to support these claims, these stereotypes took hold. Given this framing, it is not a stretch to think such stereotyping could lead to prejudicial beliefs and discriminatory actions that manifest themselves in low percentages of minorities being hired in environmental organizations.

Although environmental justice supporters have articulated an alternative frame of minorities and environment since the mid-1980s, the environmental justice counter-frame is still not the dominant frame. This is the case because activists and scholars have been slow in collecting the appropriate data to support their claims. Despite the fact that there is little evidence to support the dominant frame, supporters of alternative frames need to provide convincing evidentiary support for their counter-claims to challenge and supplant the entrenched and still-dominant frame. Although the afore-mentioned ECO study has been helpful in this regard, many years passed before any more national studies of diversity were conducted that could provide data on the status of diversity in environmental institutions. My own work is an attempt to fill that void, producing data that will provide an effective counter to the mainstream frame of minority deficiencies vis-à-vis the environmental workforce.

Recruitment is another factor that affects minority hiring in the environmental field. Relatively low percentages of students report being recruited – this despite the fact that, since the early 1990s environmental leaders have indicated that they want to diversify their organizations. In addition, environmental organizations still rely on network ties and informal mechanisms such as word of mouth to recruit their staff. Although organizations use multiple recruiting strategies, reliance on the aforemen-tioned strategies can result in filtering and homosocial reproduction of an existing workforce (in the case of mainstream environmental organizations and environmental agencies, the homosocial reproduction of predominantly white workforces).

Research has shown that social and professional networks play critical roles in finding employment, being retained, and getting promotions. This is the case because networks provide actors with weak ties that help them to maximize their social leverage and gain a competitive edge in job seeking and promotions. Networks can be seen as a form of cultural capital that can help people to bridge structural holes in their attempts to negotiate the intricacies of the workplace and attain leadership positions. That is, people who are not acquainted or connected to others in a particular organization or in a sector of the workforce are less likely to be employed or promoted to leadership positions than those who have extensive connections (Gabbay & Zuckerman, 1998; Fligstein, 1990; Burt, 1992; Granovetter, 1973, 1974; Katz & Tushman, 1981; Coleman, 1990). Social and professional networks help to build such ties. Networks also provide contacts for job seekers, timely information about jobs that may not be widely known, and an opportunity for contacts to sponsor or recommend job seekers – thereby increasing the chances of getting employment (Elliott, 2001; Aponte, 1996; Cohn & Fossett, 1996; Kasinitz & Rosenberg, 1996; Sassen, 1995; Waldinger, 1997). Researchers argue that employers favor insider (network) referrals because this an effective way of leveraging employees' social ties to the advantage of an organization. In a 1996 study, 37% of the organizations sampled reported that they frequently used insider referrals to recruit new workers. Employers use insider referrals because it expands the pool of job applicants. Moreover, because insider referrals have already passed through one layer of screening or filtering, employers believe that insider candidates are better qualified than non-referred candidates (i.e., employees know their reputations are at stake so they recommend only qualified candidates). Employers also believe that insider referrals result in a better fit between candidates and job. Referred candidates tend to know more about the job (they get information from their contacts) and the employer knows information (gleaned from contacts) about the candidate that are not usually placed on application forms. Employers believe that the information advantage that the referred candidates have over non-referred candidates can be parlayed into more rapid and effective on-the-job socialization. That is, referred candidates use their inside contacts to help them understand the rules and norms of the organization (Kalleberg, Knoke, Marsden, & Spaeth, 1996, p. 138; Fernandez, Castilla, & Moore, 2000, pp. 1288–1356; Elliott, 2001, pp. 401–425; Fernandez & Weinberg, 1997, pp. 883–902; Schwab, 1982, pp. 103–127; Breaugh & Mann, 1984, pp. 261–267).

However, insider referrals tend to be homophilous, hence the referrer-referral network tends to replicate the demographic characteristics of the incumbent workforce. That is, referrers tend to recommend someone of the same race and social background as themselves for jobs. This kind of homosocial reproduction is an important factor contributing to the lack of diversity of the environmental workforce. That is, networks act as filters that facilitate the selection of similar people (vis-a-vis their education, socialization, class, race, and social backgrounds) to work within particular organizations and sectors (DiMaggio & Powell, 1991; Granovetter, 1995; Braddock & McPartland, 1987; Model, 1993). Hence researchers argue that informal job matching that operate through internal referrals will often allow race to play a more prominent role in the hiring process, a fact that places minorities at a disadvantage (Holzer, 1987, 1996). This is the case because, despite the salience of networks in the hiring and retention of workers, studies have shown that Hispanic immigrants rely heavily on referral networks to obtain jobs (American-born Latinos do not rely as heavily on referral networks for jobs), but Blacks still rely heavily on formal methods of job matching, such as answering classified advertising, using public and private employment agencies, and walking in and applying for jobs (Elliott, 1999, 2001; Green, Tigges, & Diaz, 1999; Meir & Giloth, 1985; Holzer, 1987, 1996; Marx & Leicht, 1992).

Lack of ties to environmental networks could hurt minorities in the recruitment and hiring process. Studies have found that many minority students report that they do not have friends, acquaintances or role models working in environmental organizations (ECO, 1992). The limited extent of ties to environmental networks is still apparent today, where 63.4% of the minority students in this author's study reported knowing five or fewer people in environmental organizations. In comparison, 58.9% of whites said the same. The lack of network ties was most evident for blacks, more than one-fourth of whom reported not knowing anyone in environmental organizations. If such limited network contacts continue throughout their careers, minorities will be at a severe disadvantage in the recruitment and hiring process; minorities will also find it difficult to assume leadership roles in organizations without greater network ties.

Although organizations in the study recruit through newspapers and the web, less than 1% of mainstream environmental organizations and none of the government agencies recruited walk-ins. Although environmental organizations tout internships as a way to get a foot in the door, few reported recruiting from among their interns. Only 3.2% of the mainstream environmental organizations reported recruiting their interns and 3.9% recruited their volunteers; none of the government agencies reported recruiting through these mechanisms. Few of the environmental organizations or agencies reported recruiting from unemployment offices, either.

Although minority professional environmental associations promote jobs fairs as a way of accessing environmental jobs, few mainstream environmental organizations know of these fairs, and almost none recruit from them (corporations are much more likely to recruit at these events). While 5.9% of the government agencies reported recruiting at job fairs, only 0.6% of the mainstream environmental organizations recruited through this mechanism.[8] Although recruitment can play an important role in efforts to diversify institutions, environmental organizations have not used it effectively to identify and hire minorities. Less than 1% of the mainstream environmental organizations, and around 3% of government environmental agencies, reported recruiting from minority institutions. To be more effective, environmental organizations have to expand their repertoire of recruitment strategies. In addition to traditional methods of recruitment through networks, organizations have to do a better job of identifying where potential minority workers are (i.e., what disciplines they are in, what professional associations they belong to, etc.), and of learning how they search for jobs, to become more effective at recruiting them.

Environmental organizations also have to recognize what kinds of institutional barriers exist within the organizations that are hindering the recruitment, hiring and retention of minority workers. Data presented above show that diversity and equity factors are salient to students in the pipeline. However, for minorities who have ambitions of being promoted to senior staff positions, the statistics presented above are not very encouraging. The data imply that organizations have to dismantle the structures that shunt minorities into the diversity manager's post – which has limited potential for promotion to top leadership positions such as president or executive director, but appears much less likely to promote minorities to other leadership positions. In short, processes like filtering, insider referrals and tracking minorities into dead-end jobs result in indirect institutional discrimination. Although actors may not intend to discriminate with each act of filtering, network recruitment, hiring through word-of-mouth, promotions from only specific departments, etc., the net result is that systematic advantages are conferred on some, while systematic indirect discrimination is being meted out to others. The discrimination is institutional because of the extent to which these activities are embedded in the structure and functioning of the organizations.

Studies have found that, although some may resist efforts to diversify an institution, such efforts can get broad support. The key to the levels of support lies in the way in which diversity initiatives are framed and presented to the workforce. The fore-going discussion suggests that, if diversity initiatives are framed as initiatives that will benefit the whole workforce – rather than a specific subgroup – they are more likely to garner support from a wide range of people in the organization. Studies of diversity initiatives on college campus support this approach – such initiatives get more support from students when the benefits are perceived to be universal (Whitt, Edison, Pascarella, Terezini, & Nora, 2001).

The foregoing discussion indicates that questions about minority qualifications and willingness to participate in the environmental workforce amount to little more than distractions. The data presented above show that minority students participate in workforce preparation activities at similar rates to their white counterparts. They have sought out mentors and have participated in desirable career development activities like belonging to professional associations; they have also been attending and presenting at professional conferences at high rates. These activities are indicators of students who have gone above and beyond getting a good grade point average. They have already invested heavily in activities that will place them on a fast track to excelling at their careers and being ready to take on leadership challenges.

There is still much research to be done in this field to help us understand the dynamics of the status of diversity in the environmental field much better. Given the prominence of diversity in debates related to other sectors of the society, the debate over diversity in the environmental field is likely to garner even more attention in the future.

NOTES

1. The letter was sent to the following members of the "Big Ten" (later renamed the "Green Group" environmental groups: Natural Resources Defense Council, Sierra Club, Wilderness Society, National Wildlife Federation, Environmental Defense Fund, Friends of the Earth, the National Parks and Conservation Association, the National Audubon Society, and the Izaak Walton League.

2. Frame bridging is the process by which two structurally separate but ideologically compatible frames are connected to describe an issue. See Snow, Rochford, Worden, and Benford (1986) for a discussion of frame bridging processes.

3. The actual survey was conducted in 1988 but the results were not published until 1992.

4. See MELDI (2005) for further information on MELDI, the diversity summit and to get copies of conference reports, please visit: http://www.umich.edu/~meldi or http://www.sitemaker.umich.edu/meldi

5. For a more detailed discussion of the respondents' willingness to work in various environmental organizations and their salary expectations, see Taylor (2007a).

6. See Taylor (2007b) for a detailed discussion of the salience of institutional diversity to students.

7. The profiles of minority environmental professionals contained in *The paths we tread* attest to the significant role mentoring plays in the development of the careers of the people profiled (Taylor (2005).

8. See the MELDI website for a list of minority professional environmental associations (see, http://www.umich.edu/~meldi).

REFERENCES

Aguirre, A., & Turner, J. H., (1998). *American ethnicity: The dynamics and consequences of discrimination* (2nd ed., pp. 12–13). Boston: McGraw-Hill.

Allen, T. D., Burke, R. J., & McKeen, C. A. (1997). Benefits of mentoring relationships among managerial and professional women: A cautionary tale. *Journal of Vocational Behavior*, *51*, 43–57.

Aponte, R. (1996). Urban employment and the mismatch dilemma: Accounting for the immigration exception. *Social Problems, 43*(3), 268–283.

Banton, M., & Miles, R. (1988). Racism. In: E. E. Cashmore (Ed.), *Dictionary of race and ethnic relations* (p. 247). London: Routledge.

Banton, M. (1988a). Discrimination: Categorical and statistical. In: E. E. Cashmore (Ed.), *Dictionary of race and ethnic relations* (p. 79). London: Routledge.

Banton, M. (1988b). Institutional racism. In: E. E. Cashmore (Ed.), *Dictionary of race and ethnic relations* (p. 146). London: Routledge.

Best, J. (1987). Rhetoric in claims-making. *Social Problems, 34*(2), 101–121.

Bowker, J., & Leeworthy, V. (1998). Accounting for ethnicity in recreation demand: A flexible count data approach. *Journal of Leisure Research, 30*(1), 64–78.

Braddock, J. H., & McPartland, J. M. (1987). How minorities continue to be excluded from equal employment opportunities: Research on labor market and institutional barriers. *Journal of Social Issues, 43*(1), 5–39.

Brah, A. (1992). Difference, diversity and differentiation. In: J. Donald & A. Rattansi (Eds), *"Race," culture and difference* (pp. 126–145). London: Sage.

Breaugh, J. A., & Mann, R. (1984). Recruiting source effects: A test of two alternative explanations. *Journal of Occupational Psychology, 57*, 61–267.

Bryant, B., & Mohai, P. (1992). *Race and the incidence of environmental hazards: A time for discourse*. Boulder: Westview Press.

Burdge, R. J. (1969). Levels of occupational prestige and leisure activity. *Journal of Leisure Research, 1*, 262–274.

Bureau of Labor Statistics. (2005). *Occupational outlook handbook*. Washington, DC: U.S. Department of Labor.

Burke, R. J., McKeen, C. A., & McKenna, C. (1990). Sex differences and cross-sex effects on mentoring: Some preliminary data. *Psychological Reports, 67*, 1011–1023.

Burt, R. (1992). *Structural holes: The social structure of competition*. Cambridge: Harvard University Press.

Buttel, F. H., & Flinn, W. L. (1974). The structure and support for the environmental movement, 1968–1970. *Rural Sociology, 39*(1), 56–69.

Buttel, F. H., & Flinn, W. L. (1978). Social class and mass environmental beliefs: A reconsideration. *Environment and Behavior, 10*, 433–450.

Capek, S. (1993). The 'environmental justice' frame: A conceptual discussion and an application. *Social Problems, 40*(1), 5–24.

Carmichel, S., & Hamilton, C. (1967). *The politics of liberation in America*. New York: Penguin.

Cashmore, E. E. (1988). Prejudice. In: E. E. Cashmore (Ed.), *Dictionary of race and ethnic relations* (p. 22). London: Routledge.

CCAWMSETD, (2000). *Land of plenty: Diversity as America's competitive edge in science, engineering and technology* (pp. 1, 9). Report of the Congressional Commission on the Advancement of Women and Minorities in Science, Engineering and Technology Development. Washington, DC: United States Congress.

Census, (1990). Census of population and housing, characteristics of American Indians by tribe and language (p. CP3–7). Washington, DC: U.S. Bureau of the Census.

Census, (2000a). Table P-2. Race and Hispanic origin of people by median income and sex: 1947–1998. Washington, DC: U.S. Bureau of the Census. http://www.census.gov/hhes/income/histinc/p02.htm

Census. (2000b). *Census of population and housing, current population reports.* Washington, DC: U.S. Bureau of the Census.

Census. (2000c). Table A-3. Mean earnings of workers 18 years old and over, by educational attainment, race, Hispanic origin, and sex: 1975–1999. Washington, DC: U.S. Bureau of the Census. http://www.census.gov/population/socdemo/education/tableA-3.txt

Cohn, S., & Fossett, M. (1996). What spatial mismatch? The proximity of blacks to employment in Boston and Houston. *Social Forces, 75*(2), 557–572.

Coleman, J. (1990). *Foundations of social theory.* Cambridge: Harvard University Press.

Cotgrove, S., & Duff, A. (1980). Environmentalism, middle-class radicalism and politics. *Sociological Review, 28*(2), 333–351.

Craig, W. (1972). Recreational activity patterns in a small negro urban community: The role of the cultural base. *Economic Geography, 48*, 107–115.

Department of the Interior, Office of the Secretary (DOI). (1992). *Report of the secretary of the interior to the endangered species committee.* Washington, DC: Bureau of Land Management.

Devall, W. B. (1970). Conservation: An upper-middle class social movement. *A replication. Journal of Leisure Research, 2*(2), 123–126.

Dillman, D. A., & Christenson, J. A. (1972). The public value for pollution control. In: L. Taylor, N. H. Cheek & W. Burch (Eds), *Social behavior, natural resources, and the environment* (pp. 237–256). New York: Harper and Row.

DiMaggio, P. J., & Powell, W. W. (1983). The iron cage revisited: Institutional isomorphism and collective rationality in organizational fields. *American Sociological Review, 48*, 147–160.

DiMaggio, P. J., & Powell, W. W. (1991). The iron cage revisited: Institutional isomorphism and collective rationality in organizational fields. In: W. W. Powell & P. J. DiMaggio (Eds), *New institutionalism in organizational analysis* (pp. 63–82). Chicago: University of Chicago Press.

Doughty, R. W. (1975). *Feather fashions and bird preservation: A study in nature protection.* Berkeley: University of California Press.

Dreher, G. F., & Ash, R. A. (1990). A comparative study of mentoring among men and women in managerial, professional, and technical positions. *Journal of Applied Psychology, 75*, 539–546.

Dreher, G. F., & Cox, T. H. (1996). Race, gender and opportunity: A study of compensation attainment and the establishment of mentoring relations. *Journal of Applied Psychology, 81*, 297–308.

Dreher, G. F., & Chargois, J. A. (1998). Gender, mentoring experiences, and salary attainment among graduates of an historically black university. *Journal of Vocational Behavior, 53*, 401–416.

Dwyer, J. F., & Hutchison, R. (1988). *Participation and preferences for outdoor recreation by black and white Chicago households.* Chicago: U.S. Forest Service, North Central Experiment Station.

Dyer, R. (1988). White. *Screen, 29*, 44–46.

ECO, Environmental Careers Organization. (1992). *Beyond the green: Redefining and diversifying the environmental movement.* Boston: Environmental Careers Organization.

Elliott, J. R. (1999). Social isolation and labor market insulation: Network and neighborhood effects on less-educated urban workers. *The Sociological Quarterly, 40*(2), 199–216.

Elliott, J. R. (2001). Referral hiring and ethnically homogeneous jobs: How prevalent is the connection and for whom. *Social Science Research, 30*, 401–425.

Enosh, R. D., Staniforth, C. D., & Cooper, R. B. (1975). *Effects of selected socio-economic characteristics on recreation patterns in low income urban areas: Part II.* Madison: University of Wisconsin at Madison.

Eshner, E. A., & Murphy, S. E. (1997). Effects of race, gender, perceived similarity, and contact on mentor relationships. *Journal of Vocational Behavior, 50*, 460–481.

Environmental Careers Organization (ECO). (2001). *Increasing diversity in the environmental field: The report from the national roundtable on diversity in the environment* (p. 28). Boston: Environmental Careers Organization.

Fagenson, E. A. (1989). The mentor advantage: Perceived career/job experiences of protégés versus non-protégés. *Journal of Organizational Behavior, 10*, 309–320.

Faich, R. G., & Gale, R. P. (1971). The environmental movement: From recreation to politics. *Pacific Sociological Review, 14*(2), 270–287.

Fernandez, R., Castilla, E., & Moore, P. (2000). Social capital at work: Networks and employment at a phone center. *American Journal of Sociology, 105*(5), 1288–1356.

Fernandez, R., & Weinberg, N. (1997). Sifting and sorting: Personal contacts and hiring in a retail bank. *American Sociological Review, 62*, 883–902.

Ffloyd, M. (1998). Getting beyond marginality and ethnicity: The challenge of race and ethnic studies in leisure research. *Journal of Leisure Research, 30*(1), 3–22.

Fisher, W. R. (1984). Narration as a human communication paradigm: The case of public moral argument. *Communication Monographs, 51*, 1–23.

Flax, J. (1993). *Disputed subjects.* New York: Routledge.

Fligstein, N. (1990). *The transformation of corporate control.* Cambridge: Harvard University Press.

Gabbay, S. M., & Zuckerman, E. W. (1998). Social capital and opportunity in corporate R&D: The contingent effect of contact density on mobility expectations. *Social Science Research, 27*, 189–217.

Gamson, W. A. (1992). The social psychology of collective action. In: A. Morris & C. M. Mueller (Eds), *Frontiers in social movement theory* (pp. 53–76). New Haven: Yale University Press.

Gamson, W. A., & Meyer, D. S. (1996). Framing political opportunity. In: D. McAdam, J. D. McCarthy & M. N. Zald (Eds), *Comparative perspectives on social movements: Political opportunities, mobilizing structures, and cultural framings* (pp. 275–290). Cambridge: Cambridge University Press.

Granovetter, M. (1973). The strength of weak ties. *American Journal of Sociology, 78*, 1360–1380.

Granovetter, M. (1974). *Getting a job: A study of contacts and careers.* Cambridge: Harvard University Press.

Granovetter, M. (1995). Afterword 1994: Reconsiderations and a new agenda. In: M. Granovetter (Ed.), *Getting a job: A study of contacts and careers* (pp. 139–182). Chicago: University of Chicago Press.

Green, G., Tigges, L., & Diaz, P. (1999). Racial and ethnic differences in job search strategies in Atlanta, Boston, and Los Angeles. *Social Science Quarterly, 80*(2), 263–278.

Hannigan, J. A., (1995). *Environmental sociology: A social constructionist perspective* (p. 33). New York: Routledge.

Harris, A. (1983). Checks came in the mail, but the poison is still in the catfish. *Washington Post*, June 4, p. A2.

Harry, J. (1971). Work and leisure: Situational attitudes. *Pacific Sociological Review, 14*(July), 301–309.

Harry, J., Gale, R., & Hendee, J. (1969). Conservation: An upper-middle class social movement. *Journal of Leisure Research, 1*(2), 255–261.

Hartmann, L. A., & Overdevest, C. (1989). *Race and outdoor participation: State-of-the-knowledge review and theoretical perspective.* Missoula: U.S.D.A. Intermountain Forest Experiment Station.

Hendee, J. C., Gale, R. P., & Harry, J. (1969). Conservation, politics and democracy. *Journal of Soil and Water Conservation, 24*(November–December), 212–215.

Holden, C. (2004). Long hours aside, respondents say jobs offer 'as much fun as you can have. *Science, 304 #5678*(June), 1830–1837.

Holzer, H. J. (1987). Informal job search and black youth employment. *American Economic Review, 77*, 446–452.

Holzer, H. J. (1996). *What employers want: Job prospects for less-educated workers.* New York: Sage.

Ibarra, H. (1992). Homophily and differential returns: Sex differences in network structure and access in an advertising firm. *Administrative Science Quarterly, 37*, 422–447.

Ibarra, P. R., & Kitsuse, J. I. (1993). Vernacular constituents of moral discourse: An interactionist proposal for the study of social problems. In: J. A. Holstein & G. Miller (Eds), *Reconsidering social constructionism: Debates in social problems theory* (pp. 131–152). New York: Aldine de Gruyter.

Institute of Medicine. (1999). *Toward environmental justice: Research, education, and health policy needs.* Washington, DC: National Academy Press.

Johnson, C., et al. (1998). Wildland recreation in the rural South: An examination of marginality and ethnicity theory. *Journal of Leisure Research, 30*(1), 101–120.

Judd, R. W. (1997). *Common lands, common people: The origins of conservation in Northern New England.* Cambridge: Harvard University Press.

Kalleberg, A., Knoke, D., Marsden, P., & Spaeth, J. (1996). *Organizations in America: Analyzing their structures and human resource practices.* Thousand Oaks: Sage Publications.

Kasinitz, P., & Rosenberg, J. (1996). Missing the connection: Social isolation and employment on the Brooklyn waterfront. *Social Problems, 43*(2), 180–196.

Katz, J. H., & Ivey, A. E. (1992). White awareness: The frontier of racism awareness training. *Personnel and Guidance Journal, 55*, 485–488.

Katz, R., & Tushman, M. L. (1981). An investigation into the managerial roles and career paths of gatekeepers and project supervisors in a major R&D facility. *R&D Management, 11*, 103–110.

Kellert, S. R. (1984). Urban American perceptions of animals and the natural environment. *Urban Ecology, 8*, 220–222.

Kelly, J. R. (1980). Outdoor recreational participation: A comparative analysis. *Leisure Sciences, 3*(2), 129–154.

Kim, J. J. (2004). Optimistic new grads face reality of starting salaries. *The Wall Street Journal.* Online, March. www.careerjournal.com/myc/school/20040326-kim.html

Kornegay, F. A., & Warren, D. I. (1969). *A comparative study of life styles and social attitudes of middle income status of whites and negroes in Detroit.* Detroit: Detroit Urban League.

Kreger, J. (1973). Ecology and black student opinion. *The Journal of Environmental Education, 4*(3), 30–34.

LaBalme, J. (1988). Dumping on Warren County. In: B. Hall (Ed.), *Environmental politics: Lessons from the grassroots* (pp. 25–30). Durham: Institute for Southern Studies.

Lorber, J. (1994). *Paradoxes of gender*. New Haven: Yale University Press.

Lowe, G. D., Pinhey, T. K., & Grimes, M. D. (1980). Public support for environmental protection: New evidence from national surveys. *Pacific Sociological Review*, *23*(October), 423–445.

Lucal, B. (1996). Oppression and privilege: Toward a relational conceptualization of race. *Teaching Sociology*, *24*(July), 245–255.

Maniero, L. A. (1994). Getting anointed for advancement: The case of executive women. *The Academy of Management Executive*, *8*, 53–63.

Martinson, O. B., & Wilkening, E. A. (1975). A scale to measure awareness of environmental problems: Structure and correlates. Paper presented at the Midwest Sociological Society, Chicago.

Marx, J., & Leicht, K. T. (1992). Formality of recruitment to 229 jobs: Variations by race, sex, and job characteristics. *Social Science Research*, *76*, 190–196.

Maynard, C., Cooper & Gale, P. C. (1995). Alabama fish-monitoring data for the Huntsville Spring Branch-Indian Creek system released. *Environmental Compliance Update*, *3#9*(October).

McDonald, D., & McAvoy, L. (1997). Native Americans and leisure: State of the research and future directions. *Journal of Leisure Research*, *29*(2), 145–166.

McIntosh, P. (1990). White privilege: Unpacking the invisible knapsack. *Independent School*, *49*(2), 31–35.

Meeker, J. W., Woods, W. K., & Lucas, W. (1973). Red, white and black in the national parks. *North American Review*, *258*, 6–10.

Meir, R., & Giloth, R. (1985). Hispanic employment opportunities: A case of internal labor markets and weak-tied social networks. *Social Science Research*, *66*, 296–309.

MELDI. (2005). Minority Environmental Leadership Development Initiative. In: D. E. Taylor (Ed.), *Summaries of selected presentations from MELDI's national summit on diversity in the environmental field*. Ann Arbor: University of Michigan's School of Natural Resources and Environment. The report can be view at http://www.umich.edu/~meldi or http://www.sitemaker.umich.edu/meldi/conference_publications

Merton, R. K. (1949). Discrimination and the American creed. In: R. H. MacIver (Ed.), *Discrimination and national welfare* (pp. 99–126). New York: Harper and Row.

Milbrath, L. (1984). *Environmentalists: Vanguard for a new society* (pp. 17, 163). Albany: State University of New York Press.

Model, S. (1993). Ethnic economy and industry in mid-twentieth century Gotham. *Social Problems*, *44*, 445–463.

Mohai, P. (1985). Public concern and elite involvement in environmental conservation issues. *Social Science Quarterly*, *55*(4), 820–838.

Mohai, P. (2003). Dispelling old myths: African American concern for the environment. *Environment*, *45*(5), 10–26.

Mohai, P., & Bryant, B. (1998). Is there a race effect on concern for environmental quality. *Public Opinion Quarterly*, *62*(4), 475–505.

Morrison, D. E., Hornback, K. E., & Warner, W. K. (1972). The environmental movement: Some preliminary observations and predictions. In: L. Taylor, N. H. Cheek & W. Burch (Eds), *Social behavior and natural resources and the environment* (pp. 259–279). New York: Harper and Row.

Mueller, E., & Gurin, G. (1962). *Demographic and ecological changes as factors in outdoor recreation*. ORRRC report #22, General Printing Office, Washington, DC.

Municipal Archives. (1797). *Letter from residents of the sixth ward sent to City Hall.* New York: New York.

Municipal Archives. (1888). *Letter from Albert Oelzer sent to City Hall.* New York: New York.

National Association of Colleges and Employers (NACE). (2006). *Spring 2006 salary survey.* Bethlehem: NACE.

National Science Board. (2006). *Science and engineering indicators 2006* (p. NSB 0601). Arlington: National Science Foundation, Division of Science Resources Statistics.

National Science Foundation. (1996). *Women, minorities, and persons with disabilities in science and engineering: 1996.* Arlington: National Science Foundation.

National Science Foundation. (2000). *Women, minorities, and persons with disabilities in science and engineering: 2000.* Arlington: National Science Foundation.

National Science Foundation. (2001). *Employment preferences and outcomes of recent science and engineering doctorate holders in the labor market* (p. NSF 02–304). Arlington: National Science Foundation, Division of Science Resources Statistics.

National Science Foundation. (2005). *Recent engineering and computer science graduates continue to earn the highest salaries* (p. NSF 06–303). Arlington: National Science Foundation, Division of Science Resources Statistics.

National Science Foundation. (2006). *Characteristics of doctoral scientists and engineers in the United States: 2003.* Arlington: National Science Foundation.

New York Times. (1990). Earth issues lure a new breed of young worker. *New York Times,* July 20, p. 41.

Noe, F. P. (1974). Leisure life styles and social class: A trend analysis, 1900–1960. *Sociological and Social Research, 58,* 286–294.

O'Leary, J. T., & Benjamin, P. J. (1982). *Ethnic variation in leisure behavior: The Indiana case.* Lafayette: Department of Forest and Natural Resources. *Station Bulletin, 349.*

Omi, M., & Winant, H. (1993). On the theoretical status of the concept of race. In: C. McCarthy & W. Crichlow (Eds), *Race, identity and representation in education* (pp. 3–10). New York: Routledge.

Perrow, C. (1974). Is business really changing. *Organizational Dynamic, 3*(Summer), 31–44.

Press, R. M. (1981). 'Love Canal South': Alabama DDT residue will cost millions to clean up. *Christian Science Monitor,* June 12, p. 11.

Rafter, N. (1992). Claims-making and socio-cultural context in the first U.S. eugenics campaign. *Social Problems, 35,* 27.

Ragins, B. R., & McFarlin, D. (1990). Perceptions of mentor roles in cross-gender mentoring relationships. *Academy of Management Journal, 37,* 321–339.

Rose, F. (2000). *Coalitions across the class divide: Lessons from the labor, peace, and environmental movements.* Ithaca: Cornell University Press.

Sassen, S. (1995). Immigration and local labor markets. In: A. Portes (Ed.), *The economic sociology of immigration* (pp. 87–127). New York: Sage Publications.

Schwab, D. P. (1982). Recruiting and organizational participation. In: K. Rowland & G. Ferris (Eds), *Personnel Management* (pp. 103–128). Boston: Allyn and Bacon.

Shabecoff, P. (1990). Environmental groups told they are racists in hiring. *New York Times,* February 1, p. A20.

Shaull, S., & Gramann, J. H. J. (1998). The effect of cultural assimilation on the importance of family-related and nature-related recreation among Hispanic Americans. *Journal of Leisure Research, 30*(1), 47–63.

Shinew, K. J., Floyd, M. F., McGuire, F. A., & Noe, F. P. (1996). Class polarization and leisure activity preferences of African Americans: Intragroup comparisons. *Journal of Leisure Research, 28*(4), 219–232.

Snow, D. (1992). *Inside the conservation movement: Meeting the leadership challenge* (pp. 47–51, 111–114). Covelo: Island Press.

Snow, D., & Benford, R. D. (1988). Ideology, frame resonance, and participant mobilization. *International Social Movement Research, 1*, 197–217.

Snow, D. A., & Benford, R. D. (1992). Master frames and cycles of protest. In: A. Morris & C. M. Mueller (Eds), *Frontiers in social movement theory* (pp. 133–155). New Haven: Yale University Press.

Snow, D. A., Rochford, E. B., Worden, S., & Benford, R. D. (1986). Frame alignment processes, micro-mobilization and movement participation. *American Sociological Review, 51*, 464–481.

Sosik, J. J., & Godshalk, V. M. (2000). The role of gender in mentoring: Implications for diversified and homogenous mentoring relationships. *Journal of Vocational Behavior, 57*, 102–122.

Stamps, S., & Stamps, M. (1985). Race, class and leisure activities of urban residents. *Journal of Leisure Research, 17*(1), 40–56.

Stanton, R. G. (2002). *Environmental stewardship for the 21st century: Opportunities and actions for improving cultural diversity in conservation organizations and programs.* Washington, DC: Natural Resources Council of America.

Taylor, D. E. (1989). Blacks and the environment: Toward and explanation of the concern and action gap between blacks and whites. *Environment and Behavior, 21*(2), 175–205.

Taylor, D. E. (1991). *Determinants of leisure participation: Explaining the different rates of participation of African Americans, Jamaicans, Italians, and other whites in New Haven.* Doctoral dissertation, Yale University, New Haven.

Taylor, D. E. (1992). Can the environmental movement attract and maintain the support of minorities? In: B. Bryant, & P. Mohai (Eds), *Race and the incidence of environmental hazards* (pp. 28–54, 224–230). Boulder: Westview Press.

Taylor, D. E. (1997). American environmentalism: The role of race, class and gender. 1820–1995. *Race, Gender and Class, 5*(1), 16–62.

Taylor, D. E. (1999). Mobilizing for environmental justice in communities of color: An emerging profile of people of color environmental groups. In: J. Aley, W. Burch, B. Canover & D. Field (Eds), *Ecosystem management: Adaptive strategies for natural resource organizations in the 21st century* (pp. 33–67). Washington, DC: Taylor and Francis.

Taylor, D. E. (2000). The rise of the environmental justice paradigm: Injustice framing and the social construction of environmental discourses. *American Behavioral Scientist, 43*(4), 508–580.

Taylor, D. E. (2002). *Race, class, gender and environmentalism.* Seattle: U.S.D.A., Forest Service, PNW-GTR 534.

Taylor, D. E. (2005). *The paths we tread: Profiles of the careers of minority environmental professionals.* Ann Arbor: University of Michigan. See: http://www.umich.edu/~meldi

Taylor, D. E. (2007a). Employment preferences and salary expectations of students in science and engineering. *BioScience, 57*(February), 175–185.

Taylor, D. E. (2007b). Diversity and equity in environmental organizations: The salience of these factors to students. *Journal of Environmental Education, 39*(1).

Taylor, D. E. (forthcoming). Outward bound: Manliness, wealth, race and the rise of the environmental movement. 1830s–1930s.

Thomas, D. A., & Alderfer, C. P. (1989). The influences of race on career dynamics: Theory and research on minority career experiences. In: M. D. Arthur, D. T. Hall & B. S. Lawrence (Eds), *Handbook of career theory* (pp. 133–158). New York: Cambridge University Press.

Tognacci, L. N., Weigel, R. H., Wideen, M. F., & Vernon, D. T. (1972). Environmental quality: How universal is public concern. *Environment and Behavior, 4*(March), 73–86.

UCC. (1987). *Toxic waste and race in the United States.* New York: United Church of Christ.

U.S. General Accounting Office (U.S.G.A.O.). (1983). *Siting of hazardous waste landfills and their correlation with the racial and socio-economic status of surrounding communities.* Washington, DC: U.S. General Accounting Office.

van Ardsol, M. D., Sabagh, G., & Alexander, F. (1965). Reality and the perception of environmental hazards. *Journal of Health and Human Behavior, 5,* 144–153.

Van Liere, K. D., & Dunlap, R. (1980). The social bases of environmental concern: A review of hypothesis, explanations, and empirical evidence. *Public Opinion Quarterly, 44*(2), 181–197.

Waldinger, R. (1997). Black/immigrant competition re-assessed: New evidence from Los Angeles. *Sociological Perspective, 40,* 365–386.

Wallace, J. E. (2001). The benefits of mentoring for female lawyers. *Journal of Vocational Behavior, 58,* 366–391.

Warren, L. S. (1997). *The hunter's game: Poachers and conservationists in twentieth century America.* New Haven: Yale University Press.

Washburne, R. F. (1978). Black under-participation in wildland recreation: Alternative explanations. *Leisure Sciences, 1*(2), 175–189.

Washburn, R. F., & Wall, P. (1979). Cities, wild areas, and black leisure: In search of explanations for black/white differences in outdoor recreation. Paper INT-249, U.S. Forest Service, Intermountain Forest and Range Experiment Station, Ogden.

Washburne, R. F., & Wall, P. (1980). *Black-white ethnic differences in outdoor recreation.* Ogden: U. S. D. A. Intermountain Forest Range Experiment Station.

Washington Post. (1980). Town files DDT suit. *Washington Post,* March 17, p. A5.

Washington Post. (1982). Dumping on the poor. *Washington Post,* October 12, p. A12.

Wellman, D. (1977). *Portraits of white racism.* New York: Cambridge University Press.

Whitt, E. J., Edison, M. I., Pascarella, E. T., Terezini, P. T., & Nora, A. (2001). Influences on students' openness to diversity and challenge in the second and third years of college. *The Journal of Higher Education, 72*(March/April), 172–204.

Wilson, J. A., & Elman, N. S. (1990). Organizational benefits of mentoring. *Academy of Management Executive, 4,* 88–94.

Wright, S. (1975). Explaining environmental commitment: The role of social class variables. Paper presented at the Annual Meeting of the Midwest Sociological Society, Chicago.

Yancey, W. L., & Snell, J. (1971). *Parks as aspects of leisure in the inner-city: An exploratory investigation.* Nashville: Vanderbilt University.

Yu, P., & Berryman, D. (1996). The relationship among self-esteem, acculturation, and recreation participation of recently arrived Chinese immigrant adolescents. *Journal of Leisure, Research, 28*(4), 251–273.

Zinger, C. L., Dalsemer, R., & Magargle, H. (1972). Environmental volunteers in America. Prepared by the National Center for Voluntary Action for the Environmental Protection Agency, Office of Research and Monitoring, Washington, DC.

EXTENDING THE REACH OF ECOFEMINISM: A FRAMEWORK OF SOCIAL SCIENCE AND NATURAL SCIENCE CAREERS

Anne Statham and Christine Evans

ABSTRACT

This chapter examines relationships between gender equity and environmental concerns as expressed through two different views of ecofeminism, those of a natural scientist and a social scientist. Personal experiences are recorded and analyzed to show similarities and differences in life and career trajectories, in part influenced by ecofeminist thought. In tracing this impact, we observed that much of the current philosophical and social science framework is less applicable to a natural science perspective. Natural systems repeat and nest at varieties of scales; thus the connectivity within any system parallels, reflects, mirrors the connectivity of other systems. These parallel systems can be nested in fractal-like natural worlds, where connections within are reflected between, and the patterns of the system are replicated in each. Thus, when we look across the range of interconnected systems, the axes are not intersecting at all, but simply reflective parallels. Such may be the case with the axes of oppression emphasized by many ecofeminists. We thus propose an extension to ecofeminist thinking – the notion of system reflectivity that

Equity and the Environment
Research in Social Problems and Public Policy, Volume 15, 149–170
ISSN: 0196-1152/doi:10.1016/S0196-1152(07)15004-3

encompasses, but is broader than, the idea of simultaneously operating axes of oppression.

INTRODUCTION

This chapter examines relationships between gender equity and environmental concerns as expressed through two different views of ecofeminism. We use a comparative ethnographic framework to trace the impact of ecofeminist thinking on the careers of a social scientist and a natural scientist, in the process critiquing ecofeminist theory and offering extensions to this body of thought. We consider which aspects of this approach have furthered our efforts to comprehend and accurately describe reality and to communicate to our students efficacious ways of approaching their own work. We end by considering how this conversation between us has informed our individual viewpoints and created new understandings of the relationship between gender and environmental concerns, offering an extension to ecofeminist thought.

OUR UNDERSTANDINGS OF ECOFEMINISM

Ecofeminism, a body of thought built on the notion that gender and environmental issues have deep overlaps, has been defined in a variety of ways that appeal to a multidimensional audience. For example, Sturgeon (1997) lays out five variants of the idea of a "special connection" between women and nature that is at the core of most ecofeminist theories: (1) that patriarchy equates women and nature, requiring a feminist analysis to disentangle the two; (2) that women are defined as closer to nature, dooming them to an inferior position; (3) that women more readily notice environmental issues because of the traditional gender roles they are assigned (agricultural production and managing households); (4) that women are biologically closer to nature by virtue of their bodily functions – regular menstrual cycles, pregnancy, lactation, etc.; and (5) that nature-based religions contain strong images of female power, an antidote to the relatively powerless position women are assigned in most contemporary societies.

Warren and Cheney (1991) discuss ten similarities between ecological feminism and ecology, including the emphasis both approaches place on the importance of context, world views, anti-reductionism, integration of viewpoints, actual complementarity between presumed dualisms, complexity in interrelationships/scales/other dimensions, whole system behavior, network or relational views of systems, and the history of the system or process. Greta Gaard (2002) traces the uses animal rights activists have made of the theory. It is a very eclectic theory that has been used by those with a variety of viewpoints. Indeed, some argue for a more postmodern complexity in ecofeminist thought that does not reduce to any equation of women with nature (Salleh, 1997).

Sturgeon argues that ecofeminism can be seen as a feminist rebellion within radical environmentalism, and traces its development through its first public event in the U.S., the Women and Life on Earth conference at Amherst in 1980, organized by Ynestra King (Institute for Social Ecology), Anna Gyorgy (Clamshell Alliance), and Grace Paley (pacifist activist). King eventually distanced herself from the Institute for Social Ecology and Murray Bookchin's brand of that theory (Sturgeon, 1997).

Karen Warren (2002) has more recently emphasized the "contingent interconnections" among many social and economic processes that are presumed and examined within the framework of an ecofeminist approach, while continuing to focus on the interlocking axes of oppression or prism of "isms" – race, gender, class, environmental degradation, etc. However, one of her critics, Glazebrook (2002) notes that "it is not always clear what the nature of the connection is" between feminism and environmentalism.

We examine the interconnections from the standpoints of a social scientist and a natural scientist, offering some extensions of ecofeminist thought in the process, and attempting to clarify the nature of that connection. For example, many ecofeminists focus on the interlocking axes of oppression, and this forms the heart of much of Statham's work. Our conversation, however, has caused us to reconsider ecofeminist thought as a specific example of system overlap, which is a more general and powerful concept for accurately depicting the complexities of reality, helping us to see the patterns that exist.

OUR JOURNEY

We both first became interested in ecofeminism in the 1970s, when some of the first works were being published. The word "ecofeminism" was not always used, but the idea of a connection between women and nature was growing. In fact, Susan Griffin's (1978) book was called *Woman and*

Nature: The Roaring Inside Her. This work pulled together strands of issues of concern to us – feminist critique, environmental awareness, spiritual searching. And Carolyn Merchant's (1980) book, *The Death of Nature: Women, Ecology and the Scientific Revolution,* announced the prophecy carried through by many ecofeminists, namely that our social and economic orders may well result in the death of nature, if radical changes are not made. More recently, Merchant (2003) has extended that analysis, arguing that the consumer ideology and identities promulgated by Western civilization are taking us to the brink of ruin, and recounting efforts by thinkers such as Glendinning (1994) to bring us back from the brink. There are numerous examples of these approaches. Early proponents of the female-centered religion and deities approach who influenced Statham included Charlene Spretnak (1982), Nor Hall (1980), and Riane Eisler (1988).

Vandana Shiva, an Indian biologist and activist, working with several others, is perhaps most well-known for her arguments that women are most directly affected by environmental destruction, or most likely to notice, by virtue of their concerns for family. With Maria Mies, she has pointed to the impact of the worldwide destruction of subsistence economies on indigenous peoples, arguing that women and children are often last in the new line-up for resources, hence hurt most by environmental destruction. She points especially to the "General Agreement on Tariffs and Trade" (GATT) treaties and their impact on biodiversity and intellectual property, ultimately shaping the ability of indigenous people to grow their own food (Mies & Shiva, 1988). Later, with Jayanta Bandyopadhyay, she traced the impact of market economics on nature and survival, again with an eye to the impact on "efficient" indigenous societies and the role that women have played in the "front lines" of the movement (Bandyopadhyay & Shiva, 2001), touching on several of the strands outlined by Sturgeon. These works have influenced both Statham and Evans. Evans was additionally influenced at an earlier point by Starhawk's "Spiral Dance" (1988) and James Lovelock's "Gaia Hypothesis" (Lovelock, 1979, 1988), which was also promulgated by Fester and Margulis (1991).

METHOD

Our data for this chapter are our personal experiences. As with any sort of "naturalistic inquiry" (Lincoln & Guba, 1985), we have grappled with the need to impose some sort of order and consistency on an essentially unwieldy set of observations. There is a growing literature on efforts to analyze one's personal experiences as data. Conventionally, topics are explored, even with qualitative approaches, by observing others, interviewing them, or gathering documents such as newspaper clippings, as was done recently by Capek (2006), as she explored the idea of surface tension existing between one's self and the environment in the aftermath of a cattle egret slaughter in her neighborhood. However, some have begun to look at their own behavior and communications. For example, Vaughan (2006) used e-mails she sent and received in the aftermath of the Columbia disaster to explore how the sociological concepts she developed around the Challenger disaster "traveled" across various boundaries during the later period. Ellis (2004) has pioneered the use of personal notes and journals to write fictionalized autoethnographies that reveal the sociological principles emergent from experience.

We opted to use retrospective journals that chronicled our movement into our careers and our encounters with ecofeminism and environmental issues as connected with gender. We used these journals, along with the papers we have published, much as one would use interview transcripts in a conventional qualitative analysis. We first developed categories that represented the patterns that emerged from our reading of these journals and papers, using the constant comparison method developed by Glaser and Strauss (1967) to identify, develop, and analyze the themes discussed below. We then grouped our material into these categories and analyzed the content of each category, looking for emergent concepts and patterns.

Our analysis is done within the framework of five emergent categories: (1) learning about the issues, (2) developing a world view, (3) becoming a scientist, (4) seeing connections, and (5) doing the work. We talk first about our own individual trajectories along these lines, then look at the commonalities and differences in our experiences, with what results. Throughout this analysis, we look for both the impact of ecofeminist thought on our lives and careers, and then at what our conversation can contribute to the extension or clarification of ecofeminism.

STATHAM'S EXPERIENCE

Statham's mother nurtured a perceived connection with nature. This happened through outings, picnics, etc., that they did as a nuclear family (often without her father, who claimed that his time in the army had spoiled any love for camping he might have had) and also with her grandparents. Statham's mother was very active in Girl Scouting and sent her, her sister and brother off to scout activities and camp during the summer. Girl Scouting was an avenue for developing a sense of being connected to the environment in some of the women she studied later, as well.

Because of this early connection with nature, she became interested in environmental issues while in college – along with her interest in peace issues, civil rights, and feminism. This was in the late 1960s; she graduated in 1970. She joined several environmental organizations, and with her 1991–92 sabbatical, she was able to bring this avocation into her professional repertoire. After that sabbatical, she began teaching an environmental sociology course, and spent a summer teaching this course and an ecofeminism graduate seminar at the University of Wisconsin in Madison.

The first of the five positions outlined by Sturgeon has appealed most to Statham. In attempting to disentangle the connection drawn between women and nature, an emerging idea, increasingly seen in feminist writing, was that of interlocking axes of oppression, that is, the idea that our economic system requires groups of expendable labor, for profit maximization, a need that in time produced a corresponding system of exploitation of certain groups – by race, ethnicity, gender. The argument was that systematic discrimination against members of certain groups served the needs of our economy, and that it is not possible to understand fully how one of these factors operates without considering the total context. Certain ecofeminists had also begun to argue that nature was treated in ways similar to women – as a resource to be exploited for the maximization of profit in our economic system – and further that women's future is tightly tied to the future of the environment.

After being trained as a quantitative researcher at a leading Midwestern University, Statham left her first job and went to a place where she could retool, and begin doing qualitative research. This gave her time to read and write about women's issues from a less positivistic perspective than she had felt forced to adopt during her graduate training. She began doing community-based research with her students and on her own. She became Director of the Women's Studies Program, and later the Program Administrator for the University of Wisconsin System's Women's Studies Outreach Program. Shortly before assuming this last position, she took a sabbatical and went to South Florida to study the viewpoints of those doing environmental work, compared with the viewpoints of those invested in other issues – feminism, race/ethnic discrimination, and economic development. She was interested in how the individuals involved constructed these issues as connected or distinct. How were the "axes of oppression" – argued to exist by ecofeminist theory – seen in the ways in which individuals thought about and constructed their behaviors and sense of self? Did they perceive these issues as connected at root? How does awareness of the "interlocking axes of oppression" play out in individual consciousness, and how did this awareness make coalition politics more or less possible?

The things she has published on environmental awareness and gender came from this participant observation and depth interview study conducted during her sabbatical granted for the 1991–1992 academic year. She did participant observation and depth interviews with 33 participants in four groups involved in various issues – environmental restoration, feminist issues, race and class issues, and economic development – exploring commitment to specific issues and perceived overlap among the various issues. Nearly a quarter of her sample was African American or Native American, and nearly half was male.

Two main, relevant points emerged from those publications. First, in a study done for *NWSA Journal* (Statham, 2000), she found ecologists to have a better sense of how these various issues connect, and to see more clearly the connected axes of oppression supposedly at the core of the problems posited by ecofeminists. The feminists and others she studied tended to see distinctions among these issues, and in fact often saw them as competing – for funding and attention. A telling example was the comment that a feminist made to her:

> Greenpeace has tons of money, they don't need mine ... [they] can publish one picture of a baby seal and ... bring in $10 million ... . The feminist movement has had a much more difficult time fund raising.

A possible explanation for this difference in seeing connection or competition among these issues, Statham hypothesized, was the disjunction feminists had experienced in their philosophical development. They had been raised to believe one thing about gender roles, then had this notion drastically uprooted in adolescence or adulthood. The environmentalists, by contrast, had been nurtured in their world views very consistently by parents, especially mothers, and they were able to live out these early beliefs, often much more fully then their parents had been able to do. Perhaps this sense of continuity carried over into their constructions of issue connections, as well. However, it seemed the ecological worldview itself, which stresses the interconnections of all things, was also a factor here. Whatever the reason, her hope was that insights gleaned from this comparison would spur feminists reading the NWSA Journal in which the article was published to rethink their approaches to single issues, and to broaden their support for related issues held dear by groups they hoped to form coalitions with – enhancing chances for successful coalition politics. It seemed that those who were aware of the interconnections between issues, the axes of oppression discussed by ecofeminists, had the potential of engaging in more effective coalition politics. In this chapter, she had recounted a situation where members of a feminist organization had been invited by Native American women to join them protesting the landing of the Columbus ship replicas in Miami, and the feminist group had arrived with their Keep Abortion Legal signs, which greatly offended the Native American women. To say the least, this destroyed their chances of working collaboratively in a sustained way, since many women of color in our society have a history of sterilization promotion by the U.S. government, and are hence wary of (if not downright hostile toward) birth control programs.

In an earlier article Statham (1995) had explored the ideas of "connected" sense of self and "integrated" sense of self. In this paper, her intent was to suggest advances to aspects of traditional identity theory suggested by ecofeminist principles. First, traditional identity theory tends to see the "self" as a closed system, whereby an individual maneuvers to maintain a sense of self established fairly early in the socialization process. A closed system is posited through which the individual takes in information from the outside, but tends to interpret it and then react based on longstanding notions of priorities, salience, and ultimately, significant others' expectations. Evidence suggests that even those with a deviant or "spoiled" sense of self are very resistant to new, corrective input. (Stryker & Serpe, 1994; Stryker & Burke, 2000).

However, there are alternative systems of thought – such as those from Eastern or indigenous perspectives – that suggest the possibilities for a sense of self more closely and continuously connected to the community and/or the environment, and hence, less of a closed system. Evidence from Statham's study suggests that there are degrees of this more connected sense of self within our own culture, as well, something not well researched, even at the present time. However, perhaps this tendency to see oneself as more closely connected to others and environment is a disadvantage in a society with a market economy, where individual success is so much rewarded. On the other hand, changing all of these tendencies is a priority urged by ecofeminist thinkers discussed earlier in this chapter, such as Glendinning.

Second, traditional identity theory tends to posit fairly separate sub-identities that correspond to roles played, that comprise an individual's sense of self. Statham's data suggest that for certain groups, for example the environmentalists, these separate identities may be seen as more integrated than presumed in the traditional identity models. By incorporating these ideas, identity theorists could broaden their repertoire of tools to account more fully for the importance of connections to nature and community in individuals' efforts to construct and maintain their sense of self in this postmodern world (Gergen, 1991; Gillard & Laudine, 2006). In a third, still-unpublished paper (Statham & Gann, 2006), Statham explores the factors that determine one's sense of connection to the immediate physical environment, and the role that being a native – born and raised in a place – plays in this development.

Her teaching has also been shaped by this move toward environmental issues. As she began to teach the environmental sociology course, she was becoming convinced that learning was greatly enhanced by building community projects into courses, especially with a course like this, with so many ways to apply what is being learned. As a result, she has always taught this course with some community project incorporated, once traveling back to South Florida and providing opportunities for students to engage in environmental restoration activities in a variety of settings. In this way, students have become her collaborators, some of them forming long-term relationships with her, as they explore ways to build environmental action into their careers and lives more generally. Statham's research and teaching programs have been directly influenced by her interest in ecofeminism. The pedagogy she has used, which emphasizes connections and direct experience, has also been influenced by this body of thought.

EVANS' EXPERIENCES

The 1970s were certainly Evans' time of becoming acquainted with feminism, as well. The connection to ecofeminism, however, was much later and separate, signifying her initial linear/traditional approach to scholarly endeavors. She also had early experiences of connecting to nature with hikes that she and her brother enjoyed on Saturday mornings with their little beagle, Betsy, and lunch, hunting knives, and snuck matches for a campfire, on which they would roast apples on sticks. They had a realm of "wilderness" within about two miles of their house – on the outer edge of burgeoning suburban development. Friends of her father's had a hog farm in southwestern Michigan, and she and her brother got to vacation there for a week most summers, where they enjoyed walking along the dusty road in the evening, watching fireflies and listening to the night sounds.

Her introductions to issues of feminism were later. Returning to college as a former hippie/dropout, her first response to feminism was skeptical. She disliked the idea of forming "camps" and did not initially understand that these "camps" were already formed, and were not an artifact of feminism, but an artifact of a patriarchal society. Once she understood that, she was immediately enraged at being assigned to the lesser "camp," and thoroughly embraced the "personal is political" approach.

At that time, however, she was still an outsider to science – pursuing a scientific discipline primarily because her sixties integrity forbade that she officially pursue studies she felt she could teach herself. So her resistance to credentialing became her way into the sciences, and she struggled with the quantitative, linear approach required by the discipline. In fact, she had never previously realized that one could officially "scholarize" nature, and this revelation became an exciting prism. Science was hard until she was able to see it as a way of looking at the universe that made sense to her. When she started to see that science was about relationships and connections, it became a lens on the world that she embraced. She especially appreciated that it had both breadth and metaphorical properties, even though her mostly male professors discouraged the metaphorical connections. Nonetheless, science, especially earth sciences, became her larger viewpoint. To this day, she urges her students to consider soils as "a metaphor and meditation on universal truth." They sometimes look at her oddly.

Initially, though, she saw very few existential connections between the environment and feminism. Her first views were more system-related: women had fewer opportunities in the sciences, largely because science was perceived as a "male" domain. She remembers specifically two events. First, her academic (female) advisor urged her to pursue a career directed toward writing about science/journalism vs. science per se, because there was just no sense in competing with men in "their" territory, and instead, it would be better to accept a secondary role that would involve much less conflict and strife. She was not impressed!

Second, when she applied for a summer internship with the Natural Resource Conservation Service (formerly the Soil Conservation Service (SCS)), she was advised that they would not accept women into these positions – which were career-conditional – because the SCS required travel and relocation of their employees. She was a married woman, and the agency saw it as inconceivable that her husband would agree to such disruption! Outraged, but still trying her best to "do science" in the traditional (male) way, she continued to view the issues as separate.

In the course of her doctoral work, in her late thirties, she finally began to appreciate non-linearity. Taking a metaphorical approach to the body of earth science literature, and a metaphysical approach to the "nature of nature," a blending began. It was difficult at first to see that the cyclical, webbish approach to the female nature applied, indeed, to all of nature. Blending the cycles with the linearity was an extreme challenge. And at some point, she could see that the weblike approach of feminism was entirely appropriate to the viewing and understanding of the physical world.

During that time Evans also read Starhawk's *Spiral Dance* and James Lovelock's *Gaia Hypothesis*, along with supporting work by Lynn Margulis. All emphasized intentional interconnectedness – between women and nature, within nature itself, within women's lives. Awareness and consciousness are most valued – and reliable – when they follow the circular, globe-like, or interwoven pathways. She began to consider the metaphysical more seriously: organically, spiritually, and within the definitive connections of the natural world.

Three outcomes were particularly revealing. First, Evans was drawn primarily into the surficial processes aspect of soil and earth sciences. This sub-discipline not only allows for, but requires, a multi-scale, multiple factor, and iterative approach to the chemistry of the landscape system. There is also more room for the notion of telling a credible story, which integrates the quantitative and the conceptual. The "problem" requires understanding, and an applied "solution" is not critical. In fact, practical applications may even (regrettably now) be disdained; knowledge is valued for knowledge's sake, and the multiple perspectives required allow for multiple dimensions of perception.

Second, this approach leads to a focus on connections and interactions of processes and mechanisms. This was often in conflict with some of the more stochastic approaches employed by Evans' male colleagues in the natural resource areas. Later, Barbara McClintock's biography, *A Feeling for the Organism* (Keller, 1983) enabled her to feel more supported in valuing intuitive approaches to formulating scientific questions, as did the John Nash biography, *A Beautiful Mind* (Nasar, 1998). These validated the creativity to be found within science – most notably, that the questions one asked were more important and useful than the conclusions drawn. The belief that models are most useful when they generate more good questions was not accepted by many science colleagues who sought the predictive model as the determinative outcome of scientific investigation. However, as long as investigations were carried out within an acceptable methodological framework, and provided that the initial intuitive questions could be suitably re-phrased as hypotheses, one could "do science" in a more non-linear fashion. Examples of her work that incorporate this notion are her studies of soils and stratigraphic sections and landscapes (Morton, Evans, Harbottle, & Estes, 2001, Evans, Morton & Harbottle, 1997; Langley-Turnbaugh & Evans, 1994; Evans & Roth, 1992). The focal point in these studies was to construct a representational and semi-quantitative aggregate that could be used to categorize and understand these processes. In mentoring graduate students, particularly at the University of New Hampshire, Evans was frequently at pains to re-educate her students about the value and limitations of statistical analysis to drawing inferences about process, rather than simply system status.

Reflections on Gender and Science (Keller, 1985), and various writings from Vandana Shiva, introduced the suggestion that women may naturally gravitate toward a less linear approach to science. Non-linear, "chaos theory" work in the sciences, particularly physics, also provided the fractal as a representative of complex systems. Though not specifically feminist, chaos theory of complex systems validated (for many male scientists) a non-linear perception of natural and physical systems' dynamics. For feminist scientists, the fractal – the picture within the picture, endlessly replicating – seemed simply to make sense as the proper way to query the cosmos (Gleick, 1988). Although initially bewildered by this circular approach, Evans' students began to find the "aha! moments" in their studies. One graduate student commented that whenever she began to dig a soil pit, it was like opening a present! Increasingly valuing these responses, Evans began to move toward a more integrated program of teaching, research, and service; the traditional empire building and publication of esoteric papers began to look much more trivial than it had in the past. She became involved in a difficult tenure and promotion case of a female chemist in an otherwise all-male department.

Acting on these responses, Evans was thrilled to find a place for herself at University of Wisconsin-Parkside, where the integration was similarly prized, multi-disciplinary collegiality was supported, and female faculty were significantly empowered. She felt she could be a part of something larger than herself, contributing to a reordering of her individual department (Geosciences), which she was hired to Chair, making the department an active component of the campus mission, also contributing to a larger regional mission, infusing environmental awareness into the entire process. As with Statham, her research and teaching program have also been influenced by her affinity for the ecofeminism perspective, perhaps more in terms of the processes she has used and studied than in the content of her studies and courses.

Shortly after arriving on campus, she met Statham, who has an office down the hall from hers, offering her a chance to work on civic engagement grant Statham had recently received from the U. S. Department of Housing and Urban Development. Over time, this work led to their collaborating on projects, involving students in their respective courses in the same community projects.

OUR CONVERSATION

Backgrounds

There are many similarities in the biographies of Statham and Evans. Both had early experiences of making connection with nature, Statham through her mother's influence and that of Scouting, Evans through her outdoor activities with her brother and later with other friends. Thus they early on experienced the complexity and force of nature directly, which led to an interest in environmental issues for both of them. Statham also had early experiences with issues of race and class that she later began to integrate for herself when she was in college in the late 1960s.

Both Statham and Evans rebelled against controlling fathers. In somewhat different ways, their relationships with their fathers entailed their first memories of issues of gender, and the constraints they would later face, although neither seemed to have defined things in that way at the time. Both fathers claimed the gendered boundaries their daughters faced were the product of the world "out there," but did not explicitly say their daughters were not capable of male-typed activities.

Both Statham and Evans took explicit stances against gender boundaries during the late 1960s and early 1970s, while they were in college. Statham had also become interested in environmental issues, as well as race, class, and anti-war activities. While she did not articulate for herself that these issues were connected, working on them simultaneously began to foster that belief. During college, she transformed from a typical coed into something of a campus radical. When she was first married, right out of college, she went with her then-husband to a military base where he was stationed, and this caused her to think more explicitly about her place in these issues and her commitment to such causes as the anti-war movement. While there, she taught General Equivalency Diploma (GED) classes and later, in her hometown, worked for the welfare department. As such, she continued to encounter issues of race and class and gender and to forge a viewpoint of commitment to addressing the various "isms" emphasized by much ecofeminist thinking.

During this period, Evans became aware of the inequalities experienced by women, and became a feminist. Her efforts to get an internship with the Natural Resource Conservation Service were blocked because she was a married woman. This enraged her, but she did not at that point see the connections between the issues of gender and the environment. "Feminism is about justice, and science is about facts," she recalls thinking.

Both experienced issues of entering male-dominated situations while in graduate school. Statham recalls a classmate in her graduate program yelling at her husband at a crowded party, after she had received one of four National Institutes of Health (NIH) quantitative fellowships midway through her first year of graduate school, "What does she do when you start making love to her? Spew out numbers?" Evans recalls a visit by an Equal Employment Opportunity (EEO) representative from the United States Department of Agriculture (USDA) visiting a project she was working on who praised her for "being able to dig holes and core trees as well as the men."

Both came to see science as a viewpoint they could use to pursue their goals, in somewhat modified ways, and each adopted stances somewhat at odds with the male-dominated paradigms surrounding them. In a new job, Statham learned qualitative methods and in her current job, began using that perspective almost exclusively, finding it more satisfying in examining the issues that truly interested her. She also began building a model for collaborative research in a long-term, state-wide research project she has led on the issue of women and poverty and welfare reform. This and her prior work on how individuals experience the "isms" emphasized by ecofeminists as connected, and how efforts to address them (coalition politics) are furthered if participants can see the connections, was for her a way of using the scientific approach in way that was satisfying to her.

Evans also struggled with the quantitative linear approach required by her discipline, but eventually developed a model that stresses the relationships and connections inherent in the scientific viewpoint. As an example, Evans cultivated a particularly integrated approach to the celebration of Ground Hog Day. Imbolg, "in the belly" (of the mother) is the Wiccan festival that celebrates the first stirrings of spring as the midpoint between the winter solstice and the spring equinox, occurring on February 1. In many of these rituals, the remnants of the preceding season's grain are retrieved, blessed, and planted as the seeds of the next season's harvest. The following day, western popular culture also turns attention to a creature within the earth – the ground hog – whose response to the warming and increase of light prompt a venture from hibernation. In the Christian calendar, February 3 is Candlemas, the 40th day following Christmas, traditionally the postpartum purification. Symbols of the waxing light in this ritual include blessing and lighting of candles. As these celebrations of increasing light and earth awakening occur immediately around Evans' birthday (Feb. 1), she has adopted the Ground Hog as a special totem animal appropriate to her earth-centered sciences.

As she pursued her metaphorical approach to the "nature of nature," she began to see that the "webbish" approach to the female nature applied to all of nature, and was an appropriate way to understand the physical world. Thus, she began to see the connections between the issues of gender and environment. After reading Starhawk and Lovelock, then Margulis, she began to see the "circular, globe-like, interwoven pathways...between women and nature, within nature itself, within women's lives." For her, awareness and consciousness became a priority.

Doing the Work

Perhaps because they were both working against the grain of their respective disciplines in many ways, both came to emphasize the importance of doing meaningful work as somewhat distinct from the common notions of success in their disciplines. As Evans put it, earlier in this paper, "I began to move toward a more integrated program of teaching, research, and service; the traditional empire building and publishing of esoteric papers began to look much more trivial than it had in the past." Familiar networks for her are natural systems, and her own perception was that a more intuitive comprehension of connectivity is probably an "essential" characteristic of women.

Statham finds that she enjoys knowing that her work may have some impact on the "real world," and is not only for colleagues' consumption. She especially enjoys mentoring students in community projects, some of which have had much impact. She has developed a Community-Based Learning program for the campus. However, she faces the tension between scholarship and activism that traditional notions of academia entail, especially in the social sciences, where boundaries between these two endeavors are actively patrolled.

They both believe these projects are a direct result of the paradigmatic approach they have embraced, which emphasizes connectivity and complexity in the world they are attempting to depict. In this model, getting the "right answer" quickly is less important than thoroughly exploring the "messiness" or chaos of the reality being studied. Rejection of dualisms and understanding balance among forces is also a key feature of the approach they both use, in their research and in their teaching. As several ecofeminists have contended, emphasizing the interconnectedness of the natural and social world enhances the possibilities for seeing what is "really" there (i.e., Sandilands, 1999).

Their collaboration has also influenced their viewpoints. They have collaborated on several such projects in their Environmental Studies courses, where their students work on the same project, emphasizing different aspects – the natural science component or the social ramifications. Students have benefited enormously from seeing both perspectives simultaneously, developing their critical thinking skills and becoming less concerned about "the right answer." These experiences will also benefit them outside of the classroom, in showing them the value of cooperation, discussion and interaction, and giving them a less isolated sense of self in viewing the rest of the world.

In these courses, Evans and Statham took an environmental justice perspective, something else they discovered they had in common. Taylor (1997) argues that ecofeminists have tended to ignore issues of concern to women of color, but Statham and Evans, in drawing on the work of those such as Vandana Shiva, find interconnections between the two approaches, specifically the notion that race can play a factor in exposure to environmental hazards – both domestically and worldwide (see also Chapter 3 by Taylor, this volume). Others have argued convincingly that this is so (Bullard, 1993, 1990; Frey, 2001). While some controversy exists about this claim (Anderton, 1996; Lambert, Boerner, & Clegg, 1996), other evidence bolsters these original claims (Mohai, 1995; see also especially the chapters by Mohai, this volume; Bullard, this volume).

In their work on this issue, students find a concrete way to see the interconnections of issues and systems. One project dealt with lead safety. Students in both courses were forced to understand the science of lead contamination within the context of the socio-economic factors that influenced production of the hazard and then vulnerability to exposure among different groups in our society. In this case, Evans' passion for environmental justice actually pushes her to step outside of her self-consciously neutral role and attempt to become an outside observer or agent, involved in a cause. Interestingly, though, when she queries her students about the concept of environmental justice, they most often believe that it means "doing" justice to the environment, i.e., treating it fairly and respectfully. Sometimes they believe that environmental justice is simply the retaliation of the environment, trading poisoning for pollution, or hurricanes for greenhouse gases, in an eye-for-an-eye fashion. An appalling number of her students seem to believe that most of "those people" should just try harder to move up in life. This is, of course, an extreme expression of the outsider viewpoint.

Evans believes that feminism is, above all, a vision of inclusivity. Similarly, her view that nature does not contain "we's" and "they's" prohibits alienation from any natural occurrence, situation, or viewpoint. We cannot have objective outsider convictions because we are neither outside nor objective. Divorcing oneself from commonality is an artificial schism that ultimately stunts vision, humanity, and perhaps salvation. Thus, these collaborative course projects can better broaden her students' viewpoints than any other type of experience.

Statham faces somewhat different issues. Her students are very aware of the unequal position of many citizens in our society, and given their largely working-class backgrounds, they are not surprised to learn that certain stigmatized groups suffer more from environmental degradation. However, they tend to feel very powerless about making a difference, and they need much encouragement to see how the efforts they can make – and do make in these projects – can have an impact in the larger scheme of things. Thus, their interactions with Evans and her students gives them greater hope that solutions can be found.

There are also several areas of difference that have emerged from their interactions. First, in tracing the impact of the ecofeminist perspective on their life and career trajectories, they have come to see that much of the current philosophical and social science framework is less applicable to a natural science perspective. The emphasis on the axes of oppression is a case in point. Natural scientists are less concerned with this than with understanding how the natural world works. Ecofeminists have helped them to see the connections between those two realms, but as Salleh earlier stated, the actual connection is not always clear. Perhaps a consideration of the "overlap of systems" would be a metaphor to add to ecofeminist tools, as a more general way of discussing the axes of oppression. It would allow us to think of it in broader terms.

Incorporating some aspects of earth systems science into ecofeminist theory could broaden the audience able to use the theory. A model within earth system science asserts connections within given systems are more similar than different. Thus, systems tend to mirror each other, rather than being tightly coupled. Natural systems repeat and nest at varieties of scales;

thus the connectivity within any system parallels, reflects, and mirrors the connectivity of other systems. Systems are linked by similarity. In examining a fractal approach to objectivity, the self becomes a component of multiple systems; the more detailed the range of scale of examination, the more thorough the understanding of both "self" and "system." Using "self" as a reference point, rather than seeking objectivity, results in multiple expressions of belonging. Also, parallel systems can be nested in fractal-like natural worlds, where connections within are reflected between, and the patterns of the system are replicated in each.

Thus, when we look across the range of interconnected systems, the axes are not intersecting at all, but simply reflective parallels. Such may be the case with the axes of oppression emphasized by many ecofeminists. By looking at various levels and directions of connections, the individual may be shown ways to gain more power in resisting the situation. Personal power comes from being a self-aware part of as many systems as possible. The pattern of the system of systems is probably replicated somewhat in each. Having this grounding can enhance the likelihood of being effective. Understanding the individual linking mechanisms and aspects of each system is thus of utmost importance. This may advance women's strategic interests, as called for by Buckingham (2004), who suggests framing women's interests in a wider framework than the additive nature of gender analysis.

Another difference that has emerged is the way Statham and Evans define "problem." This has led to some interesting discussion, in their collaboration and in writing this paper. For Evans, a "problem" is value-neutral – simply a puzzling set of observations whose connections are not yet understood. Satisfaction comes from understanding, regardless of whether that understanding changes anything in the objective world. This may be because she does not readily disconnect herself from the objective world. Instead, she strives to remain attentively connected to as many aspects and scales as possible – from the individual organism, to her perceptions of herself as a spiritual, feminist/female scientist, teacher, citizen – experiencer of multiple cosmic environments. Her scientific stance allows her to remain connected while observing. In many ways, the attempt to be simultaneously viewing through multiple scales of perspective makes it more difficult for her to assemble a fixed viewpoint, or an absolute notion of "should" – as distinct from her own personal sense of right-feeling.

For Statham, distance from her social surroundings is often required in doing her work. She strives to see the things others miss or take for granted, to raise their awareness of the human-made nature of social patterns, urging a consideration of other options that may be healthier in the end.

This tendency is apparent in the work she discusses here – her efforts to show how coalition movements could be forged if different approaches were used, or that common notions of identity may be helping to fuel problems in seeing how we are connected to each other and to our surroundings.

This difference may be related to perspectives within the social and natural sciences on the social vs. the technical aspects of issues. The observer and the observed cannot be separated, Evans believes. Though certainly there are methodological approaches that help to more clearly define a "problem," the observer is inextricably linked to that "problem" and cannot truthfully stand outside of it to choose a single, correct solution, even if there *is* a single correct solution. This difference between them may also be related to the tension between basic and applied research within their respective disciplines.

While conventional notions of science are based on the assumption of "objective" observations, conventional science itself becomes an obstacle to holistic methodologies, at least in part because this assumption of objectivity is flawed. This has led Evans to use perspectives on problems that others have not used, or to combine viewpoints that others see as separate. For example, she has consistently varied her scales of observation, giving her a more contextual and iterative framework for the problems she has studied than the more typical monoscale approach can yield.

Social scientists also struggle with the issue of "objectivity" and with the connection between the observer and observed, striving to predicate/ prescribe an observer's viewpoint that takes subjectivity into account, while consciously attempting to maintain a distinction between the viewpoints of the observer and the observed. As a result, while Statham has used the concept of ecofeminism to explore views of oppression connected in some way to gender, Evans has used it to explore parallels and intersections of events and observations through the perspective of the particular set of lenses she chooses to use. For her, the connection between the environment and feminism are not based on a commonality of abuse or degradation, because that would require an objectively correct notion of issue and solution.

CONCLUDING COMMENTS

Writing this chapter has given us an opportunity to discover common interests we had not discovered thus far in our working relationship. We share concerns about the major issues touched upon here. We have looked at the issue of gender and the environment from several positions. Gender issues influence who is admitted into the professions of those who do research on environmental issues. Gender may influence how individuals tend to look at or experience environmental degradation. Ecofeminist principles have provided areas for Statham to research and an intellectual structure for both Evans and Statham to use in their work overall. We have found our teaching and service to be infused by ecofeminist principles of justice and inclusivity – both in terms of content and pedagogy.

Our comparative approach has revealed the various ways theoretical frameworks can find their way into academic work across a spectrum of disciplines. Articulation of our respective viewpoints will likely be fruitful in further collaborative work. We also propose an extension to ecofeminist thinking – the notion of system reflectivity that encompasses but is broader than the idea of simultaneously operating axes of oppression.

REFERENCES

Anderton, D. (1996). Methodological issues in the spatiotemporal analysis of environmental equity. *Sociological Quarterly, 77*, 508–515.

Bandyopadhyay, J., & Shiva, V. (2001). Development, poverty, and the growth of the green movement in India. In: S. Frey (Ed.), *The environment and society reader*. Needham Heights, MA: Allyn and Bacon.

Buckingham, S. (2004). Ecofeminism in the twenty-first century. *The Geographical Journal, 170*, 146–154.

Bullard, R. (1990). *Dumping on Dixie: Race, class, and environmental quality*. Boulder, CO: Westview Press.

Bullard, R. (1993). *Confronting environmental racism: Voices from the grassroots*. Boston, MA: South End Press.

Bullard, R. (this volume). Equity, unnatural man-made disasters, and race: Why environmental justice matters. In: *Research in social problems and public policy* (Vol. 15).

Capek, S. (2006). Surface tension: Boundary negotiations around self, society, and nature in a community debate over wildlife. *Symbolic Interaction, 29*, 157–181.

Eisler, R. (1988). *The chalice and the blade*. New York: HarperCollins.

Ellis, C. (2004). *The ethnographic I: A methodological novel about autoethnography*. Newbury Park, CA: AltaMira Press.

Evans, C. V., Morton, L. S., & Harbottle, G. (1997). Pedologic assessment of radionuclide distributions: Use of a radio-pedogenic index. *Soil Science Society of America Journal, 61*(5), 1440–1449.

Evans, C. V., & Roth, D. C. (1992). Conceptual and statistical modes to characterize soil materials, landforms, and processes. *Soil Science Society of America Journal, 56*(1), 214–219.

Fester, R., & Margulis, L. (Eds). (1991). *Symbiosis as a source of evolutionary innovation: Speciation and morphogenesis*. Cambridge, MA: MIT Press.

Frey, S. (2001). The hazardous waste stream in the world-system. In: S. Frey (Ed.), *The environment and society reader*. Needham Heights, MA: Allyn and Bacon.

Gaard, G. (2002). Vegetarian ecofeminism. *Frontiers, 23*, 117–146.

Gergen, K. (1991). *The saturated self*. New York: Basic Books.

Gillard, P., & Laudine, C. (2006). Maintaining the connections that sustain community-with each other and the world of nature. Paper for the Second International Conference on Environmental, Cultural, Economic and Social Sustainability. Hanoi and Ha Long Bay, Vietnam.

Glaser, B., & Strauss, A. (1967). *The discovery of grounded theory*. New York: Aldine Publishing.

Glazebrook, T. (2002). Karen Warren's ecofeminism. *Ethics and Environment, 7*, 12–26.

Gleick, J. (1988). *Chaos: Making a new science*. New York: Penguin.

Glendinning, C. (1994). *My name is Chellis and I'm in recovery from Western civilization*. Boston: Shambala.

Griffin, S. (1978). *Woman and nature: The roaring inside her*. New York: Harper and Row.

Hall, N. (1980). *The moon and the virgin*. New York: Harper and Row.

Keller, E. F. (1983). *A feeling for the organism the life and work of Barbara McClintock*. San Francisco, CA: W.H. Freeman.

Keller, E. F. (1985). *Reflections on gender and science*. New Haven, CT: N Yale University Press.

Lambert, T., Boerner, C., & Clegg, R. (1996). *A critique of "environmental justice"*. Washington, DC: National Legal Center for the Public Interest.

Langley-Turnbaugh, S. J., & Evans, C. V. (1994). A determinative soil development index for pedo-stratigraphic studies. *Geoderma, 61*, 39–59.

Lincoln, Y. S., & Guba, E. G. (1985). *Naturalistic inquiry*. Newbury Park, CA: Sage.

Lovelock, J. (1979). *A new look at life on earth*. New York: Oxford University Press.

Lovelock, J. (1988). *The ages of Gaia: A biography of our living earth*. New York: W. W. Norton.

Merchant, C. (1980). *The death of nature: Women, ecology and the scientific revolution*. San Francisco, CA: Harper and Row.

Merchant, C. (2003). *Reinventing Eden: The fate of nature in Western culture*. New York: Taylor and Francis.

Mies, M., & Shiva, V. (1988). *Ecofeminism*. London: Zed Books.

Mohai, P. (1995). Methodological issues: The demographics of dumping revisited: Examining the impact of alternate methodologies in environmental justice research. *Virginia Environmental Law Journal, 13*, 615–653.

Mohai, P. (this volume). Equity and the environmental justice debate. In: *Research in social problems and public policy* (Vol. 15).

Morton, L. S., Evans, C. V., Harbottle, G., & Estes, G. O. (2001). Pedogenic fractionation and bioavailability of uranium and thorium in naturally radioactive spodosols. *Soil Science Society of America Journal, 65*(4), 1197–1203.

Nasar, S. (1998). *A beautiful mind. A biography of John Forbes Nash, Jr*. New York: Simon and Schuster.

Salleh, A. (1997). *Eco-feminism as politics: Nature, Marx, and the postmodern*. New York: Zed Books.

Sandilands, C. (1999). *The good-natured feminist: Ecofeminism and the quest for democracy*. Minneapolis: University of Minnesota Press.

Spretnak, C. (1982). *The politics of women's spirituality: Essays on the rise of spiritual power within the feminist movement.* New York: Anchor Books.

Starhawk. (1988). *The spiral dance: A rebirth of the ancient religion of the great goddess.* San Francisco, CA: Harper and Row.

Statham, A. (1995). Environmental identity: Symbols in cultural change. *Studies in Symbolic Interaction, 17,* 207–240.

Statham, A. (2000). Environmental awareness and feminist progress. *NWSA Journal, 12,* 91–105.

Statham, A., & Gann, G. (2006). A new way of relating to the environment: Does it matter if you were born there? Unpublished paper. University of Wisconsin-Parkside, Kenosha, Wisconsin.

Stryker, S., & Burke, P. (2000). The past, present, and future of an identity theory. *Social Psychology Quarterly, 63,* 284–297.

Stryker, S., & Serpe, R. (1994). Identity salience and psychological centrality: Equivalent, overlapping, or complementary concepts? *Social Psychology Quarterly, 57,* 16–35.

Sturgeon, N. (1997). *Ecofeminist natures: Race, gender, feminist theory, and political action.* New York: Rutledge.

Taylor, D. (1997). Women of color, justice, and ecofeminism. In: K. Warren (Ed.), *Ecofeminism: Women, culture, nature.* Bloomington, IN: Indiana University Press.

Taylor, D. (this volume). Diversity and the environment: Myth-making and the status of minorities in the field. In: *Research in social problems and public policy* (Vol. 15).

Vaughan, D. (2006). NASA revisited: Theory, analogy, and public sociology. *American Journal of Sociology, 112,* 353–394.

Warren, K. (2002). Response to my critics. *Ethics and the Environment, 7,* 40–59.

Warren, K., & Cheney, J. (1991). Ecological feminism and ecosystem ecology. *Hypatia, 6,* 179–197.

PART III:
EQUITY AND INEQUALITY IN
EXTREME ENVIRONMENTS

IN THE SHADE OF AFFLUENCE: THE INEQUITABLE DISTRIBUTION OF THE URBAN HEAT ISLAND

Sharon L. Harlan, Anthony J. Brazel,
G. Darrel Jenerette, Nancy S. Jones, Larissa Larsen,
Lela Prashad and William L. Stefanov

ABSTRACT

The urban heat island is an unintended consequence of humans building upon rural and native landscapes. We hypothesized that variations in vegetation and land use patterns across an urbanizing regional landscape would produce a temperature distribution that was spatially heterogeneous and correlated with the social characteristics of urban neighborhoods. Using biophysical and social data scaled to conform to US census geography, we found that affluent whites were more likely to live in vegetated and less climatically stressed neighborhoods than low-income Latinos in Phoenix, Arizona. Affluent neighborhoods had cooler summer temperatures that reduced exposure to outdoor heat-related health risks, especially during a heat wave period. In addition to being warmer, poorer neighborhoods lacked critical resources in their physical and social environments to help them cope with extreme heat. Increased average temperatures due to climate change are expected to exacerbate the impacts of urban heat islands.

Equity and the Environment
Research in Social Problems and Public Policy, Volume 15, 173–202
Copyright © 2008 by Elsevier Ltd.
ISSN: 0196-1152/doi:10.1016/S0196-1152(07)15005-5

INTRODUCTION

Within the last few years, the subject of climate change has been widely discussed in the pages of scientific journals (e.g., Patz, Campbell-Lendrum, Holloway, & Foley, 2005; Easterling et al., 2000; Kalkstein & Greene, 1997), weighty international reports (IPPC, 2007; Watson & the Core Writing Team, 2002; WHO, 1990), a wave of popularizing books (e.g., Gelbspan, 2004; Speth, 2004), and articles in nature and news magazines (e.g., Appenzeller, Dimick, Glick, Montaigne, & Morell, 2004; Kluger, 2006). Climate change is even the subject of two feature films, *The Day After Tomorrow (2004)* and *An Inconvenient Truth (2006)*. There is good reason to concern ourselves with the earth's climate, because historically, climate change has been a huge influence on the fates of human civilizations, determining where they thrive and when they fail (Redman, 1999). In contrast to natural variations in past eras, however, it is now a virtual certainty that social (anthropogenic) forces have been driving global warming during at least the last 50 years (IPPC, 2007). Today the social, economic, and political implications of potential climate change appear to be enormous (Easterling et al., 2000; King, 2004; Motavalli (2004); Patz et al., 2005; IPCC, 2001).

Less publicized than global change, but perhaps even more immediately threatening to most people's everyday lives, is regional climate change, caused by large-scale, rapid urbanization. Cities are becoming warmer (Oke, 1997; Brazel, Selover, Vose, & Heisler, 2000), in some cases much warmer, with disastrous consequences for some people. Each summer, cities all over the world experience "perfect storm" weather conditions for extreme heat events, because they combine and concentrate in large population centers the effects of the "urban heat island" (UHI), episodic heat waves, and a warming world (NYCHP, 2004; McGeehin & Mirabelli, 2001; ICLEI, 1998).

International policy organizations, medical and social researchers, and environmental justice activists have identified health equity issues in human exposure to extreme temperatures. Entire regions, nations, and major segments of the population, defined by economic status, race, age, and disability, are vulnerable to climate-related mortality and morbidity (e.g., CBCF, 2004; Basu & Samet, 2002; Semenza et al., 1996). A few studies have also evaluated community characteristics associated with human vulnerability to extreme heat (Browning, Wallace, Feinberg, & Cagney, 2006; O'Neill, Zanobetti, & Schwartz, 2003; Klinenberg, 2002).

Our study expanded these perspectives on health equity by investigating the relationships among microclimates, vegetation and land use patterns, and the socioeconomic characteristics of people who live in different neighborhoods in the Phoenix, Arizona metropolitan region. Phoenix is located in the desert of southwestern US and is widely known for both the positive and negative aspects of its climate – a comfortable vacation playground for the well-to-do in the winter and a searing dry heat for permanent residents in the summer. We asked who is more vulnerable to extreme heat during the summer in Phoenix because of the places where they live and the resources they have to cope with excessively warm weather. We believe that urban climate systems[1] are an environmental justice issue, because poor and minority people are more likely than affluent white people to live in climatically stressed neighborhoods. And we also believe that people in poor neighborhoods are more vulnerable to extreme heat – meaning that they are at greater risk of experiencing heat-related illness – because they have fewer material and social resources to cope with extreme heat.

THEORETICAL PERSPECTIVES ON URBAN SOCIOECOLOGICAL SYSTEMS

Urban ecology focuses on the physical environments of cities and how their biophysical attributes (climate, biodiversity, hydrology, soil, and the like) differ and change compared to uninhabited places (Grimm, Grove, Pickett, & Redman, 2000). An important, relatively new idea emerging from this line of inquiry is that human management of urban ecosystems has produced greater spatial heterogeneity in the configuration of environmental resources, such as species abundance and diversity, when compared to undeveloped areas (Faeth, Warren, Shochat, & Marussich, 2005; Martin, Warren, & Kinzig, 2004; Hope et al., 2003; Pickett et al., 2001). The new urban ecology incorporates social drivers, such as the economy, polity, and culture, into conceptual modeling of environmental change and the resulting feedback effects to people's quality of life (Grimm et al., 2000; Redman, Grove, & Kuby, 2004; Pickett et al., 2001; Collins et al., 2000). However, the social drivers of change and the mechanisms of environmental resource allocation remain at a highly abstract conceptual level in ecological theories.

Social theories of urbanization and stratification are critical to understanding complex socioecological systems. Land use provides an important linkage with urban ecology. Questions of how land is used, who owns what land and where, and what happens to land cover in the process of urbanization – all may codetermine the spatial distribution of human settlement patterns and environmental resources in cities. Nearly a century ago, human ecologists recognized that the physical structure of cities is the result of economic competition for land and that where people live is determined by their social position (Park, 1915; Burgess, 1928). Contemporary political ecology underscores that urban space is divided by powerful economic and political elites who gain control of land in order to shape and manage social and physical environments to their advantage (Harvey, 1973; Kirkby, O'Keefe, & Howorth, 2001; Bryant, 1998).

Increasingly, land in developing areas of the world, as well as in the US, is highly commodified, privatized and controlled by development corporations with the acquiescence of governments (Logan & Molotch, 1987). Socioeconomic elites own the best land, reaping the rewards of local environmental amenities, such as clean air, safe drinking water, biodiversity, open spaces, and shade. On the other hand, limited economic resources, political disenfranchisement, and legally enforced restriction to undesirable land, burdens marginal social groups with degraded landscapes lacking in physical amenities that are distributed through human management of environmental resources. There are strong associations between degraded land, the existence of environmental hazards (or lack of amenities) and the location of marginalized populations (Blaike, Cannon, Davis, & Wisner, 1994).

The residential segregation and stratification of urban neighborhoods by social class and race/ethnicity provides a context in which to examine systematically the degree to which spatial heterogeneities in biophysical variables observed by ecologists are coterminous with social patterns of human settlement. Residential segregation is an important part of contemporary urban social structure that leads to persistent inequalities (Massey & Denton, 1993). Despite some decline in African American segregation between 1990 and 2000, major urban centers in the US continue to be highly racially segregated (Sethi & Somanathan, 2004). Hispanic segregation shows an increase (Iceland, 2004; Logan, Stults, & Farley, 2004), and social class segregation has increased due in part to the congregation of affluent households in specific municipalities within metropolitan areas (Fischer, Stockmayer, Stiles, & Hout, 2004). The social consequences of segregation for marginalized poor and minority groups are well-documented in terms of inferior educational and job opportunities (Kozol, 2005; Wilson, 1987), high crime rates (Sampson, Raudenbush, & Earls, 1997), health risks (Schulz, Williams, Israel, & Lempert, 2003), and industrial hazards (Pellow, 2000).

Our research is also related to urban spatial stratification, but it is focused specifically on human-induced inequalities in the biophysical environment. The climate of Phoenix is, in fact, altered by local human activities. We hypothesized that variation in human-managed vegetation and land use patterns across the urban area would produce a temperature distribution that is spatially heterogeneous and correlated with the social characteristics of urban neighborhoods in much the same way that other social and environmental inequalities are distributed among advantaged and disadvantaged neighborhoods.

THE HEAT ISLAND: AN URBAN SOCIAL PROBLEM

Understanding human impacts on regional climate begins with an explanation of how the UHI is formed as an unintended consequence of humans building upon rural and native landscapes to create cities. As cities grow, the buildings and vast expanses of paved parking lots and roads made of asphalt and concrete increase heat-absorption and heat-storage capacity (Oke, 1997; Voogt, 2002). These impervious surfaces release heat more slowly at night than natural ground covers. Other sources of anthropogenic heat (e.g., vehicles, air conditioners, and industry) exhaust heat into the air and produce emissions with contaminants that trap heat close to the ground (Grossman-Clarke, Zehnder, Stefanov, Liu, & Zoldak, 2005). Urban vegetation also plays a role in altering shade and rates of evapotranspiration. These combined processes produce higher nighttime temperatures and generally higher but more variable daytime temperatures in cities than in nearby suburban and rural areas. Cities situated in many types of climate regimes have heat islands (Patz et al., 2005). A schematic representation in Fig. 1 shows the UHI effect for a desert city, such as Phoenix, where it is mainly observed as higher nighttime temperatures. Average maximum and minimum temperatures vary widely over different types of land uses in the metropolitan area.

The UHI exacerbates the severity of heat waves as critical temperature thresholds are crossed more readily and more frequently. Heat waves – abnormally elevated temperatures over a period of days – cause more deaths worldwide than all other extreme weather events combined (CDC, 2004; NOAA, 2005; Larsen, 2006; IFRC, 2003; Curriero et al., 2002; Smoyer, Rainham, & Hewko, 2000). Future projections indicate that warmer weather will continue, the frequency of severe weather events will increase, and city residents will be most at risk for the consequences (NYCHP, 2004; Kalkstein and Greene, 1997). With nearly half the world's population living in urbanized areas, and most of the projected future growth to occur in cities (Cohen, 2003), urban climate systems are of great importance to humanity.

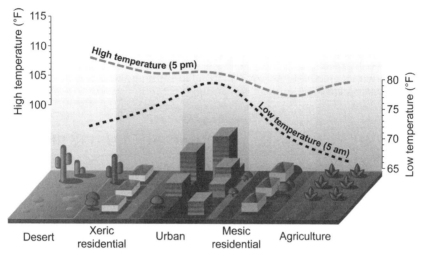

Fig. 1. Temperature Variation with Land Use Change in an Urban Environment. Xeric landscaping uses individually drip-watered, drought tolerant plants on a crushed stone groundcover. Mesic landscaping uses a mixture of high water-use vegetation and shade trees with turf grass (Martin et al., 2003). *Source:* Joseph Zehnder.

METHODS

Overview of the Project

Our Neighborhood Ecosystems project was an interdisciplinary study of the impact of urbanization on human–ecological–climate interactions. Investigators from five disciplines – sociology, geography, ecology, geological sciences, and urban planning – worked closely together in order to advance understanding of coupled human-natural systems in the Phoenix region. First, we identified the concept of "neighborhood" as a spatial unit that organizes elemental social, ecological, and climate systems (e.g., Sampson, Morenoff, & Gannon-Rowley, 2002; Pickett & Cadenasso, 1995; Addicott et al., 1987). Second, using data available from satellite imagery that allowed us to see the landscape of urbanized areas in new ways (Liverman, Moran, Rindfuss, & Stern, 1998), as well as portable climate monitoring stations, we observed that there are different microclimates in neighborhoods where privileged and underprivileged people live in the region. Third, we integrated socioeconomic, vegetation, land use, and temperature data in models that translated these differences into unequal exposure to heat-related health risks for residents of eight neighborhoods. Although our research did not go far enough to demonstrate how local microclimates are altered by real estate markets, city zoning, economic resources, and consumer taste, we did suggest plausible explanations and social implications for spatial variation in vegetation and temperature.

Phoenix, Arizona – Summer in the Desert

It will probably surprise the native, and surely the stranger, who has heard tell of Phoenix heat, to know that the normal summer temperature of Phoenix is 86.7 degrees. *The Arizona Graphic*, 1899

Today's resident or summer visitor would be even more surprised than her ancestor to know that in 1899 the Salt River, cottonwood trees, pastures, and open spaces made Phoenix a place of relatively moderate outdoor temperatures in the summer, especially between dusk and dawn. Fig. 2 compares the same corner of downtown Phoenix in the 1870s and 2007 to show how the landscape has changed. Established in the 1860s by Anglo settlers as an agricultural community, Phoenix has been residentially segregated by ethnicity and race since the 1890s, when Mexicans and other groups began to arrive in numbers (Bolin, Grineski, & Collins, 2005). During the 20th century, Phoenix (population 5,000 in the 1890s) grew to the sixth largest metropolis in the United States, encompassing 24 municipalities with a population of 3.5 million in 2005. In 2000, approximately 30 percent of the population was Hispanic, 8 percent was other minorities, and 62 percent was Anglo. Residential segregation is still prevalent and the conversion of agricultural fields, ranches, and pristine desert land into residential developments advances steadily. The river that once flowed through downtown Phoenix and provided irrigation for native vegetation and agriculture, has been dammed far upstream, leaving a dry riverbed in the city. The daily average summer temperature had risen to 95°F (35°C) by 2003. Growth is projected to continue unabated until the population reaches 8 million in 25 years.

The average annual minimum temperature rose much faster in the city than in the rural areas of Maricopa County during the last century. The rural–urban comparison in Fig. 3 shows that a minimum temperature difference of 4°F (2.2°C) in 1949 became a 9°F (5°C) difference in 2006. Urban temperatures track regional urban population growth, and Phoenix has one of the fastest warming rates in the world for its population size

Fig. 2. Washington Street, Phoenix in the 1870s (Top) (Courtesy of Herb and
Dorothy McLaughlin Photograph Collection, Arizona State University Libraries).
Same Corner at Washington Street and First Avenue in 2007 (Bottom) Photograph
by Darren Ruddell.

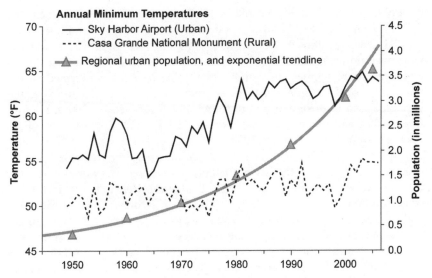

Fig. 3. Rapid Urbanization and Rising Temperature in Phoenix, Annual Minimum Temperatures in Phoenix vs. Rural Area. *Source:* Anthony Brazel.

(Brazel et al., 2000). The Phoenix region is exceptionally warm all summer, and like metropolitan areas in more temperate climate zones in the U.S. and abroad, Phoenix experiences summer heat waves that cause deaths and heat-related illnesses. More than 1,200 heat-related health incidents in the region were reported to local fire departments in each of the last five years (Golden, 2006). Ironically, the exponential growth in Phoenix that has contributed to the heat island was catalyzed by the invention of refrigeration that enabled people to cool indoor temperatures with air conditioning during the summer months (Luckingham, 1989).

Data Sources

Data for the Neighborhood Ecosystems project were collected regionally over an area of approximately 925 mi^2 (2,396 km^2) and for eight particular neighborhoods. Regionally, we combined remotely sensed satellite imagery of vegetation abundance and surface temperature, population characteristics from the 2000 U.S. Census, and a digital elevation model of the region's topography. Vegetation abundance, measured by the Soil-Adjusted Vegetation Index (SAVI) (Huete, 1988), and kinetic surface temperature were obtained from data collected by the Enhanced Thematic Mapper Plus (ETM+) onboard the Landsat 7 satellite during passes over Phoenix on May 21, 2000 and June 9, 2001 around 11:00 a.m. local time. These data sets were overlaid using a Geographic Information System (GIS) and scaled so that vegetation, temperature, and topographic patterns conformed to the boundaries of the region's 634 census tracts. Technical explanations of the variables used in our analyses are found in Jenerette et al. (2007), Harlan, Brazel, Prashad, Stefanov, and Larsen (2006), and Stefanov, Prashad, Eisinger, Brazel, and Harlan (2004).

The locations of the eight neighborhoods we studied in greater detail were selected from 46 sites in the city of Phoenix where ecologists working with the Central Arizona-Phoenix Long-Term Ecological Research (CAP LTER) project periodically collect field data (Grimm & Redman, 2004). Originally, the neighborhoods were selected to represent different types of communities for the purpose of conducting a social survey (Harlan et al., 2003; Larsen et al., 2004). Neighborhoods were defined by census block group boundaries (which are smaller and more homogeneous areas than census tracts), superimposed around the 9,688 ft^2 (900 m^2) CAP LTER sites. These neighborhoods varied in median income, ethnic composition, age of housing stock, types of landscaping, and locations in the urban core, suburban areas, and the fringe of new development. The neighborhoods were arrayed on a 30-mile (48 km) arc that extends from the northern fringe of settlement through the central city and down to the southern fringe.

In addition to conducting a survey of 302 total households in these neighborhoods, we acquired data on housing quality for survey respondents, interviewed key informants in two of the neighborhoods, and installed portable air temperature/dew point loggers for 12 consecutive months in the backyard of one residence in each neighborhood (Fig. 4). The selected yards were in areas that we had previously identified through remote sensing as having average surface temperature for the neighborhoods (Stefanov et al., 2004). We created data sets for each neighborhood that included a continuous 12-month record of air temperature and humidity, U.S. Census 2000 socioeconomic characteristics, land use and land cover classifications, as well as the other household data aggregated to the neighborhood level.

Data Analysis and Synthesis

Measures of surface temperature from remote sensing and air temperature from portable monitors were each used in some of the following analyses.

Fig. 4. Portable Weather Stations in a Xeric Backyard (Top) and a Mesic Backyard (Bottom).

Surface temperature data offer the advantage of being able to compare differences across large territories, but air temperatures measured in backyards are more indicative of what people actually feel. The data were analyzed using GIS mapping software and statistical procedures, including correlations, regressions, path analysis, and analysis of variance (ANOVA).

For the eight neighborhoods, we also used qualitative assessments of the sites and a computer simulation model of human thermal comfort called OUTCOMES – OUTdoor COMfort Expert System (Heisler & Wang, 2002). OUTCOMES estimates a Human Thermal Comfort Index (HTCI) based on the energy balance of a hypothetical person, given the weather data from a site and the site's surrounding solar and thermal radiative environmental fluxes. The HTCI is a measure of heat-stress on the human body; it indicates people's level of risk for experiencing heat-related illness and mortality based on their immediate outdoor environmental surroundings. A score of 200 on the HTCI was determined to represent the "danger" zone for likelihood that people would experience symptoms of severe heat-related illness when they are outdoors. Weather data from each neighborhood, including air temperature, humidity, and other climate-related variables in the neighborhoods, such as shade, were used to model and compare levels of human physiological response to microclimate conditions. In this article, we summarize results pertaining to the summer (June, July, and August) of 2003 (Harlan et al., 2006). The results presented below draw on several published works, incorporate some previously unpublished material, and summarize the major findings related to climate inequality in Phoenix that are most relevant to social scientists.

SUMMARY OF RESULTS

Spatial differences in summer daytime temperatures detected across the entire Phoenix metropolitan region, and at the finer scale of individual neighborhoods, were significantly associated with the social class and ethnic composition of the local population. Surface temperature measured over all 634 census tracts in the region, and air temperature measured in the sample of eight neighborhood block groups, had significant negative correlations with median annual household income and significant positive correlations with percentage of Latino residents. At the regional scale, the Pearson correlations were -0.37 and 0.32, respectively; at the finer scale of neighborhoods, the correlations were -0.68 and 0.69, respectively. Considering that the zero-order correlations between temperature and these socioeconomic variables were much higher than correlations of temperature with some physical drivers of climate, such as elevation and slope, our findings suggest a strong connection between social variables and microclimates.

To illustrate the regional findings, Fig. 5 shows a shaded relief image of the Phoenix region overlaid with a June 9, 2001 daytime surface temperature scene for the census tracts in the highest quartile of the median income distribution (top) and in the lowest quartile of the income distribution (bottom). Visual inspection supports the correlations reported above: wealthier areas had lower daytime surface temperatures than poor areas. The highest-income tracts were, by and large, on the urban fringe. Lower-income tracts were mostly near the center of the city, with the main exception being the Native American communities, which are the larger darker areas to the east and south of the city in the bottom map. A path analysis controlling simultaneously for several tract-level social, demographic, and topographic variables showed that median household income and ethnic composition had relatively larger net impacts than population density and location of the tracts (measured by median year homes were built) on the spatial distribution of summer daytime surface temperatures in the region (Jenerette et al., 2007). This runs counter to the prevailing hypothesis of many climatologists that population density is the main human driver of local climate. Our findings suggest that social stratification is an important correlate of local temperature variations. Using bivariate linear regression analysis, we estimated that mean surface temperature of census tracts decreased by 0.5°F (0.28°C) for every $10,000 increase in median income for a summer day in Phoenix ($p < 0.0001$; Jenerette et al., 2007). In other words, affluent people "buy" more favorable microclimates.

In our regional analysis, population density and median income were correlated (−0.40) with each other at the census tract level – indicating not surprisingly, that affluent areas are less densely settled. Population density does play an important role in determining microclimates because it reduces temperature variation within tracts. As density increases, there are fewer possible land use configurations, and so the local environment becomes increasingly homogeneous in materials and microclimates. Thus, to live in an area with high mean temperature and low temperature variation (meaning there is nowhere to seek relief) is the worst possible microclimate for people (Jenerette et al., 2007).

In our study of eight neighborhoods, in which we measured microclimatic conditions in yards in the summer of 2003, the correlations of air temperature with income and ethnicity were even stronger than in the

HIGHEST QUARTILE OF MEDIAN ANNUAL HOUSEHOLD INCOME ($60,820 – $174,840)

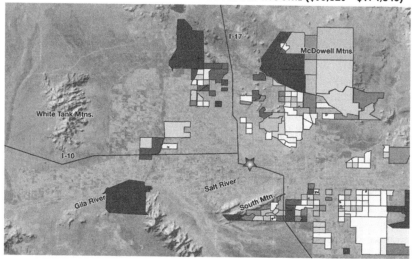

LOWEST QUARTILE OF MEDIAN ANNUAL HOUSEHOLD INCOME (< $32,733)

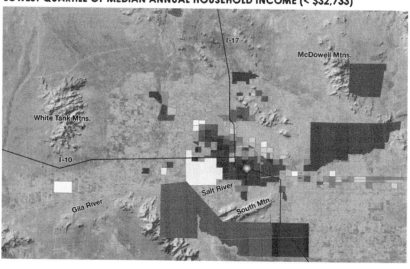

Surface Temperature in F°

96.8   109.4   116.6

Miles   10

Downtown Phoenix

N

Fig. 5. Maps of Average Surface Temperature for Census Tracts in the Phoenix Metropolitan Area: Highest Quartile of Tract Median Annual Household Income (Top) and Lowest Quartile of Tract Median Annual Household Income. (Bottom). Data derived from Landsat ETM + surface kinetic temperature for June 9, 2001 at 10:40 a.m., Maricopa County, AZ, USA. Date were processed at the Geological Remote Sensing Laboratory, Arizona State University. Map created by Lela Prashad.

regional study, as noted above. For example, during a summer heat wave in 2003, average daily air temperature at 5:00 p.m. varied by as much as 13.5°F (7.5°C) (F-test = 22.98) between an upper-income and a lower-income neighborhood located 2.5 miles (4 km) from each other in the city of Phoenix. On a "normal" summer day, the maximum difference of average daily temperature at 5:00 p.m. between neighborhoods was 7.2°F (4.0°C) (F-test = 16.39) (Harlan et al., 2006).

Vegetation abundance was the key mediating variable between median income and temperature at both regional and neighborhood scales. Changes in amounts and types of vegetation made by people during the urbanization process contribute to altering the average temperature of the region and the distribution of temperatures within the region. Temperature variation in all kinds of climate regimes can be traced to variation in land cover, even at very small spatial scales. In fact, the most direct link between people and local climate is through human-initiated changes in land cover. Anyone who has stepped from underneath the shade of a tree into full sunlight or from grass onto pavement notices this immediately.

In the regional analysis, we quantified the two-step relationship between household income, vegetation abundance (measured by SAVI), and surface temperature at the tract level. First, we found that vegetation had the largest correlation of any variable with temperature (−0.77) and the largest total effect on temperature in the path model (Fig. 6). Second, the direct positive effect of median income on vegetation was larger than the effect of any other variable in the path model, and all of the effect of median household income and half the effect of percent Latino on temperature were mediated by differences in vegetation between tracts. In other words, higher-income and predominantly Anglo tracts had greater vegetation abundance, and vegetation, in turn, predicted average surface temperature with a high degree of accuracy ($R^2 = 0.68$ for the path model in Fig. 6).

In our eight neighborhoods, vegetation abundance was also highly negatively correlated with air temperature (−0.56). During the heat wave period, this correlation became stronger (−0.70), demonstrating the importance of vegetation in mitigating heat during the times of maximum climate stress. Residents in the greenest neighborhood, Historic Anglo Phoenix, lived on palm tree-lined streets with grassy yards, shade trees, and many other plants, as well as a green park nearby (Fig. 7). Barren lots, unpaved alleys, an Interstate highway, and to be sure, a few trees, constituted much of the landscape in Black Canyon Freeway, the poorest neighborhood (Fig. 7). (For the record, Historic Anglo Phoenix consumed nearly twice as much water per household than any of the other neighborhoods.)

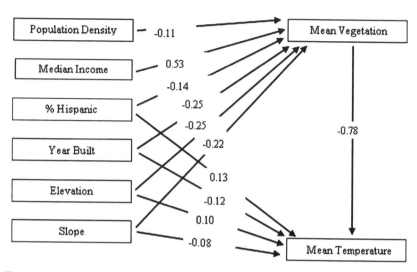

Fig. 6. Path Analysis Showing the Determinants of Vegetation and Mean Surface Temperature of 634 Census Tracts in the Phoenix Metropolitan Region. Partial regression coefficients are included on each arrow with a significant effect ($p < 0.05$). *Source:* Jenerette et al. (2007).

Our findings about the effect of income on vegetation are relevant to an emerging view of neighborhood ecology and social class, which has been labeled the "ecology of inequality" (Massey, 1996) to indicate a new ecological regime of spatially concentrated and segregated affluence and poverty in a growing urbanized population. Not only are social resources distributed unequally, but also the wealthy are privileged by having a disproportionate share of biophysical resources. For example, Hope et al. (2003) have suggested the "luxury effect hypothesis" to explain the relationship between wealth and perennial plant diversity in Phoenix neighborhoods. The authors found that vegetation diversity in census tracts with above-average income was twice that found in tracts with below-average income. In a Baltimore study, "lifestyle" variables were among the best predictors of vegetation cover in private yards (Grove et al., 2006).

We believe there are two main stratification processes at work that create class and ethnic spatial inequalities in vegetation, which we summarize briefly here. First, developers build more expensive homes on desirable lands, and these are predominantly located on the fringe of the urbanized area. This means in the Phoenix region that the most affluent neighborhoods tend to be located near agricultural fields, which are the most highly vegetated,

Fig. 7. Types of Phoenix Neighborhoods: Historic Anglo Phoenix, a High-Income, Mesic Urban Core Neighborhood (Upper Left); Black Canyon Freeway, a Low-Income, Urban Core Neighborhood (Upper Right); South Mountain Preserve, a High-Income, Xeric Urban Fringe Neighborhood (Lower Left); New Tract Development, a Middle-Income, Xeric Urban Fringe Neighborhood (Lower Right).

irrigated, and coolest places in the region. Higher-income urban fringe neighborhoods, such as North Desert Ranch (Fig. 7), offer temperature advantages because of their higher elevation, native vegetation, and rapid cooling of desert surfaces at night. Settlement patterns evident in the census data appear to indicate that people who have the means to choose, elect to live near these types of landscapes for the variety of natural amenities that they offer. Other landscape factors that contribute to cooler temperatures in wealthier neighborhoods include less dense settlement, more open space, and higher elevations.

Second, household income has a major influence on how people select and manage the vegetation in their yards, which we can demonstrate using our own results and other studies. Residential landscapes constitute a large and growing proportion of urban land cover (Lubowski, Vesterby, Bucholtz, Baez, & Roberts, 2006) and for societies of Western European heritage, the traditional lawn is imbued with cultural meaning, social prestige, and psychological gratification, as well as serving recreational and entertainment interests (Larsen & Harlan, 2006; Syme, Shao, Murni, & Campbell, 2004; Veblen, 1899). Despite the environmental excesses caused by maintaining lush greenery in the desert (due to water, fertilizer, and pesticide applications), three recent surveys found that 70–75 percent of Phoenix residents prefer "green" residential landscapes that contain at least some grass and trees over landscapes with only desert plants (Larsen & Harlan, 2006; Martin, Peterson, & Stabler, 2003; Martin, 2001; Yabiku et al., in press). A majority preferred lawn (grass and deciduous trees) or oasis (tropical plants with some grass) landscapes over desert (native plants and crushed stone) landscapes in both front and backyards. Wealthier households were significantly more likely than others to prefer the fashionable oasis landscape and to actually have the oasis, which contains a much richer array of non-native plants (including tropical ones) and denser plantings than other landscape types, as well as requiring frequent irrigation (Larsen & Harlan, 2006).

We suggest that higher-income households have the financial means to purchase and maintain more vegetation, to expand the composition of plants to more exotic species, and to cultivate their landscapes more intensively, through increasing fertilization and irrigation, in order to produce more biomass (Jenerette et al., 2007). Thus, a combination of having the means to choose homes near natural vegetation, and to manage their residential landscapes, are principal stratification mechanisms that allow wealthier households actually to create more favorable microclimates in their neighborhoods.

As a result of differences in neighborhood temperatures, exposure to heat-related health risks was significantly higher in the summer for lower socioeconomic and minority groups, especially during periods of heat waves. Because the lower socioeconomic and predominantly minority neighborhoods were significantly warmer and had other less desirable microclimate conditions, these areas had greater and longer exposure to excessive summer heat. The two lower-income Latino neighborhoods and a middle-income neighborhood on the urban fringe had the highest average HTCI scores measured at 5:00 p.m. for summer 2003. For example, most neighborhoods had average index scores at 5:00 p.m. that approached the "danger" threshold of 200 HTCI for possible sunstroke, heat cramps, and heat exhaustion. The three warmest neighborhoods, however, averaged a score of 195 and the other five neighborhoods averaged only 171. This difference intensified during a five-day heat wave in July. There was little variation among neighborhoods at 5:00 a.m. during the summer, when the average

HTCI in all was approximately equal to the lowest threshold of discomfort (meaning that people were fairly comfortable outdoors at that time of day). However, continual hourly measures of HTCI throughout the summer indicated wide disparities between neighborhoods in total time that outdoor air temperatures were above the danger threshold for heat stress. The two warmest neighborhoods had five times the amount of exposure as the coolest neighborhood (20 percent of hours compared to 4 percent of hours) during that three-month summer period.

An important part of reducing the vulnerability of high-risk populations to exposure to environmental hazards, such as extreme temperatures, is having access to resources that help people cope with the hazard (Blaike et al., 1994). In this regard, as well, the two low-income, minority neighborhoods at highest risk of outdoor heat exposure were deficient. In these neighborhoods, we found the least access to heat mitigation strategies: fewer homes with central air conditioning, fewer backyard swimming pools, and weaker self-reported social ties between neighbors. The negative correlations between these variables and HTCI for eight neighborhoods were statistically significant (-0.85, -0.83, and -0.71, respectively). Social survey results indicated that the low-income neighborhoods had less bonding social capital (trust and networks among neighbors) and less bridging social capital in that residents were less likely to take action on neighborhood environmental problems (Larsen et al., 2004).

A recently built middle-income neighborhood on the urban fringe of Phoenix was one of the warmest and least vegetated sites in our sample of neighborhoods. Some might say that the inequalities between rich and poor neighborhoods in ecological amenities, including greenery and microclimates, are not surprising. For another point of view on the familiar story of environmental injustice, however, consider New Tract Development (Fig. 7), which represents the middle-class neighborhood of the future in sprawling urban centers of the arid Southwest. In contrast to high-end housing developments on the urban fringe, where buyers pay a premium for choice locations and wide-open spaces, developers in Phoenix have been mass-producing neighborhoods for middle-income buyers, who are eager to have nicer homes and better schools for much less money than they could get closer to downtown areas. The trade-off, however, is that these neighborhoods are densely packed with homes, having little open space and only sparse vegetation around homes. In response to municipal efforts to reduce water usage, developers and homeowners associations in newer neighborhoods often restrict front yard landscaping to xeriscapes, which use drought-tolerant desert plants and decomposed crushed granite as a ground cover. If cultivated properly, xeriscape has the potential to be a more water-efficient landscape, which is important in an arid region. However, it eliminates the natural cooling effect of more lush greenery and may increase energy consumption by encouraging people to rely even more on air conditioning for comfort. With rising energy prices, refrigeration will be more costly for the middle class as well as the poor.

New Tract Development is a recently built subdivision that has the highest house-to-lot-size ratio in our sample of neighborhoods, two mobile home parks left over from earlier settlement, and a hilly, rocky butte that bisects the neighborhood. It is three times more densely settled than North Desert Ranch, a nearby upper-income neighborhood where half the land belongs to horse owners. Native vegetation in New Tract is desert scrub, but even the vacant land has a higher SAVI value than the inhabited areas, where 80 percent of the front yards are xeriscaped. The average afternoon summer temperature in New Tract was exceeded only by the warmest inner-city neighborhood, and hours of total exposure to the "danger" zone of HTCI was actually the highest of all the neighborhoods (22 percent of total summer hours).

INTERDISCIPLINARY ENVIRONMENTAL SCIENCES

This study of microclimates advanced environmental knowledge in each of the five contributing academic disciplines. In sociology, we are standing on a relatively new frontier of interdisciplinary empirical research on urban environments. Sociologists are responsible for major insights about the extent and nature of residential segregation in cities and the consequences of segregation for inequality of life chances and quality of life. Many studies of cities and communities continue to expand knowledge about stratification in urban social environment by income, ethnicity, social status, and consumer lifestyles. We chose to emphasize how some important social attributes of communities co-vary with biophysical environments that also have consequences for human health and well-being. Physical places, human modifications of biophysical properties (such as land use, vegetation, and built structures), the health risks associated with environmental quality, as well as the way people perceive the environment, are all important for understanding complex urban socioecological systems.

The field of urban planning extends beyond regulating land use to considering issues of equity, community participation, and environmental justice. While planners attempt to understand and alter the overt legacies of residential discrimination and segregation at the macro-level, they should also give more consideration to the ecological implications of future urban settlement patterns. Our finding that the heat island most severely impacts the city's poorest residents has implications for allocating city funds and regulating new development. Some of our most interesting findings point to the need for research on the complex trade-offs in individual residential decisions that involve little-understood preferences, market offerings, and investment opportunities. Identifying these trade-offs will help us to understand what desires motivate urban sprawl, how sprawl enlarges the heat island, and what initiatives might minimize these disturbances of the natural environment.

Theoretical developments in urban ecology have provided new formal conceptual models for understanding organization and change in human-dominated ecosystems. The results of our study contributed to under-standing ecological processes and patterns relevant to the key role that vegetation plays in urban ecosystems. The application of a multi-scalar approach that identified correlations between regional patterns, in-depth investigations of eight neighborhoods, and interviews with individual residents generated a unique dataset for bridging pattern and process relationships. Patterns of vegetation appear to cause cascading changes to other ecological processes and modifications to the production of additional ecosystem services, such as climate regulation, that are important to people.

Heretofore, climatologists have studied urban climate mainly by relying on data from weather stations at airports and other locations where people do not necessarily live. Although these methods are appropriate for comparing cities overall, our fieldwork showed that local neighborhood characteristics (sky view factor, moisture, vegetation, surface type) and in situ weather data relate to substantial intra-city variability in climate. In addition, remotely sensed surface temperature data linked to land cover classifications is particularly useful for improving climate and surface process (i.e. weather) models in urban environments. Our analysis of data from three remote-sensing instruments provided a common quantitative bridge between social and physical science methodologies and datasets, establishing a baseline for modeling regional and neighborhood-scale climate and human thermal comfort.

Most importantly for environmental studies, our project advanced interdisciplinary science by demonstrating that people trained in a wide array of disciplines can agree on important research questions and collaborate on investigating complex problems that require a range of knowledge and skills. We resolved issues of scale and measurement, defined key indicators, and related social and ecological variables to each other on two different spatial scales. In the future, advancement in environmental science will depend upon broadly interdisciplinary collaborations.

In conclusion, significant climate warming attributable to urbanization has been measured in the last 50 years. This is projected to continue along with urban growth and to accelerate with the additional impact of global warming. Severe weather events worldwide, such as heat waves that kill tens of thousands of people each year, are expected to increase. Just as the causes of these climate changes are complex, finding solutions is also challenging, because the environmental resources necessary to mitigate heat – land, water, and energy – are increasingly scarce and allocated unequally based on ability to pay. By showing that there are many climates in a city and that people create climates and are affected unequally by them, we hope to increase awareness that it is possible to reduce risks for the most vulnerable populations.

Continuing research on the social and ecological factors that produce microclimate variation on a small scale can give engineers, designers, municipal government, and builders the knowledge needed to target heat mitigation strategies to where the most vulnerable people live. This may take much longer to happen than we would like. Despite the Phoenix region's obvious seasonal problems with extremely high temperatures, the region's local governments have been slower than in many other cities to design and implement heat mitigation strategies. Recently the city of Phoenix has established a heat island reduction initiative, formed a task force, and is seeking information from scientists and engineers at Arizona State University's (ASU) National Center for Excellence on SMART Materials for Urban Climate and Energy. At this writing, however, the task force is considering demonstration projects on parking lot surfaces and planting trees in the downtown area.

PERSONAL REFLECTIONS – SLH

One of the most important roles of social scientists in interdisciplinary collaborations on environmental science is to keep the topic of equity on the table. As I have learned first-hand, a nuanced consideration of the human condition does not receive much attention in the normal discourse of the biophysical sciences. In scientific papers that are regarded as having a social flair, people are measured by their density; that is, population density or number per km^2. In meetings, I have heard humans (rarely called people) described mainly in dismissive ways – as being too idiosyncratic to predict anything they might do in the future, for example, or as disturbances who upset the delicate balance of ecosystem functioning by putting bad things (such as nitrogen) into the environment, and perhaps most often, as being impossible to control in experimental designs and, therefore, to be avoided whenever and wherever possible. Happily, these views are becoming antiquated outlooks on the divisions between the branches of science, as evidenced by my personal experience and more broadly by the increasing number of articles in the leading scientific journals that address social problems, such as the social drivers and social outcomes of climate change.

If one is lucky, as I was, to find scientists in other fields who care about the social world as well as the natural world, human well-being as well as ecosystem functions, and who are willing to work with us, then the possibilities for understanding how human welfare depends upon the quality of the environment (and vice versa) are greatly enhanced. To be sure, working together demands extra patience and a great deal of extra time, as everyone regresses to explaining the introductory vocabulary in their respective fields. What is a pixel? An infrared band? An ecosystem service? Evapotranspiration? Social capital? A census tract? This, of course, was only the beginning of a long dialog that helped us find our way forward in a project designed to show the intuitively obvious – poor people live in warmer places. But how much warmer, why, how do we measure it, and does it matter? As time went on, I discovered that we all speak dialects of the language we call science and that we have more in common than we have differences. My colleagues and I have made a good start at answering the questions we posed, but much more remains to be done on both the science and policy of urban climate.

Many people and programs at ASU helped us in this enterprise. The Center for Environmental Studies was established in 1967 and expanded in 2005 to the Global Institute of Sustainability. Bringing together scientists with a common environmental focus and creating opportunities for interaction and collaboration, the Center/Institute, directed by archeologist Charles Redman, is an important bridge between discipline-based departments. It has also been a magnet for attracting funding from the National Science Foundation (NSF), which has directly and indirectly supported our climate work for five years.

The NSF-funded CAP LTER project, co-directed by ecologist Nancy Grimm and Redman, began in 1998 as one of only two urban LTERs in a national network of 26 sites, most of which were originally selected for their absence of human activity. I found collaborators from other sciences through my work under the auspices of CAP LTER, where I direct the Phoenix Area Social Survey. Our Neighborhood Ecosystems project also had strong ties with the NSF-funded Integrated Graduate Education and Research Training (IGERT) in Urban Ecology program at ASU. Early in the grant cycle, co-PIs Harlan, Brazel, and Stefanov conducted a one-semester IGERT Graduate Workshop on Neighborhood Ecosystems, during which we gathered and processed data and made intellectual progress on the project. Seven IGERT students from geological sciences, sociology, geography, and planning participated and shared their expertise across disciplines.

Finally, a sabbatical fellowship in 2004–2005 at the National Center for Ecological Analysis and Synthesis at the University of California, Santa Barbara, allowed me time for further study and to work on the analyses that are summarized herein (not to mention the company of many fine young ecologists and a year in a much more pleasant climate than Tempe).

Has this project been an easy ride? Well – no. I would say that we have been tolerated but not knighted (so far) for continuing to pitch climate equity to environmental scientists. Mainstream thinking still sidelines equity as a secondary issue that is not as exciting as the "real" science questions about climate change, and may even be a distraction from confronting overall societal risks. At worst, the objectivity of researchers who pursue this line of questioning is somewhat suspect. These days, however, there are usually other people at the table who support the inclusion of equity as an important topic in environmental studies.

Publishing the results of this type of research takes extra thought in framing the issues and analyses for disciplinary journals as well as luck in the peer review process. For example, one paper was turned down at a leading journal for a variety of contradictory reasons. The subject matter editor said the content was not appropriate for an ecology journal, and although the editor-in-chief disagreed with that, she was persuaded by two reviewers that it would be risky to publish the paper. One reviewer (apparently an expert in remote sensing) thought we should revise and resubmit the paper focusing on the interpretation of surface temperature and vegetation in urban environments (a primary contribution, which should be encouraged), but that we should remove the intent to find links between the neighbors' social characteristics and surface temperature, because that is an unacceptable premise. The other reviewer, undoubtedly a social scientist, wrote an extensive critique of the paper for not including enough breadth of social theory and analysis of other social factors that affect exposure of different groups (in an ecology journal!). Another of our papers spent eight months in the review process because the editor could not find reviewers for it. Both papers are now published, but it remains a challenge to reach audiences in the various fields through the disciplinary scientific literature.

Among the brighter notes about the future of interdisciplinary collaborative environmental justice research is that the LTERs, which I perceived initially as a chilly atmosphere for social scientists, are turning the corner from "ecologists only" to a more inclusive atmosphere where social scientists are among the active participants. Important recent and planned program initiatives at NSF are encouraging interdisciplinary environmental science. The increasing numbers of young faculty members trained in interdisciplinary programs are promoting collaborations. As perhaps the most tangible indicator of the directions in which I see this field evolving, I can offer the fact that I have transferred my faculty line into an interdisciplinary school that has a focus on urban environments and is establishing a Ph.D. program in environmental social science.

NOTE

1. Meteorologists characterize climate as long-term patterns and weather as short-term events. People actually experience weather (e.g., temperature and humidity) on a daily and seasonal basis but for simplicity we use the term climate to describe the system we are studying.

ACKNOWLEDGMENTS

Neighborhood Ecosystems was supported by NSF Grant No. SES 0216281. We received support from ASU, CAP LTER (NSF DEB 97114833), and the IGERT in Urban Ecology (NSF DGE 9987612) at ASU. SLH was also supported by a fellowship at the National Center for Ecological Analysis and Synthesis, a Center funded by NSF (DEB 9421535), the University of California at Santa Barbara, and the State of California. We thank our colleagues for their contributions to this research – Ed Hackett, Amy Nelson, Bob Bolin, Diane Hope, Nancy Grimm, Gordon Heisler, Shapard Wolf, Tom Rex, Andrew Kirby, David Pijawka, Charles Redman, and Jim Reichman – and our students who used this project as a learning experience and an exercise in creative thinking – Sarah Grineski, John Parker, Matthew Lord, Chris Eisenger, Nancy Selover, Danielle Zeigler, Mark Shen, and Dale Sherwood. Co-authors Nancy Jones and Lela Prashad were students and also graduate research assistants on this project.

REFERENCES

Addicott, J. F., Aho, J. M., Antolin, M. F., Padilla, D. K., Richardson, J. S., & Soluk, D. A. (1987). Ecological neighborhoods: Scaling environmental patterns. *Oikos, 49*, 340–346.

Appenzeller, T., Dimick, D., Glick, D., Montaigne, F., & Morell, V. (2004). The heat is on. *National Geographic, 206*(September), 2–75.

Basu, R., & Samet, J. M. (2002). An exposure assessment study of ambient heat exposure in an elderly population in Baltimore, Maryland. *Environmental Health Perspectives, 110*, 1219–1224.

Blaike, P., Cannon, T., Davis, I., & Wisner, B. (1994). *At risk: Natural hazards, people's vulnerability, and disasters*. New York: Routledge.

Bolin, B., Grineski, S., & Collins, T. (2005). The geography of despair: Environmental racism and the making of South Phoenix, Arizona, USA. *Research in Human Ecology, 12*, 156–168.

Brazel, A., Selover, N., Vose, R., & Heisler, G. (2000). The tale of two climates – Baltimore and Phoenix urban LTER sites. *Climate Research, 25*, 49–65.

Browning, C. R., Wallace, D., Feinberg, S. L., & Cagney, K. (2006). Neighborhood social processes, physical conditions, and disaster-related mortality: The Chicago case of the 1995 heat wave. *American Sociological Review, 71*, 661–678.

Bryant, R. L. (1998). Power, knowledge and political ecology in the third world: A review. *Progress in Physical Geography, 22*, 79–94.

Burgess, E. W. (1928). Residential segregation in American cities. *Annals of the American Academy of Political and Social Science, 140*, 105–115.

CBCF, Congressional Black Caucus Foundation. (2004). *African Americans and climate change: An unequal burden*. Washington, DC: Redefining Progress.

CDC, Center for Disease Control. (2004). About extreme heat. Available at http://www. bt.cdc.gov/disasters/extremeheat/heat_guide.asp

Cohen, J. E. (2003). Human population: The next half century. *Science, 302*, 1172–1175.

Collins, J. P., Kinzig, A., Grimm, N. B., Fagan, W. F., Hope, D., Wu, J. G., & Borer, E. T. (2000). A new urban ecology. *American Scientist, 88*, 416–425.

Curriero, F. C., Heiner, K. S., Samet, J. M., Zeger, S. L., Strug, L., & Patz, J. A. (2002). Temperature and mortality in 11 cities of the eastern United States. *American Journal of Epidemiology, 155*, 80–87.

Easterling, D. R., Meehl, G. A., Parmesan, D., Changnon, S. A., Karl, T. R., & Mearns, L. O. (2000). Climate extremes: Observations, modeling, and impacts. *Science, 289*, 2068–2074.

Faeth, S. H., Warren, P. S., Shochat, E., & Marussich, W. (2005). Trophic dynamics in urban communities. *BioScience, 55*, 399–407.

Fischer, D. S., Stockmayer, G., Stiles, J., & Hout, M. (2004). Distinguishing the geographic levels and social dimensions of US metropolitan segregation, 1960–2000. *Demography, 41*, 1–22.

Gelbspan, R. (2004). *Boiling point: How politicians, big oil and coal, journalists, and activists have fueled the climate crisis – and what we can do to avert disaster*. New York: Basic.

Golden, J. (2006). Personal communication. National Center for Excellence on SMART Materials for Urban Climate and Energy, Arizona State University.

Grimm, N. B., Grove, J. M., Pickett, S. T. A., & Redman, C. L. (2000). Integrated approaches to long-term studies of urban ecological systems. *BioScience, 50*, 571–584.

Grimm, N. B., & Redman, C. L. (2004). Approaches to the study of urban ecosystems: The case of Central Arizona – Phoenix. *Urban Ecosystems, 7,* 199–213.

Grossman-Clarke, S., Zehnder, J. A., Stefanov, W. L., Liu, Y., & Zoldak, M. A. (2005). Urban modifications in a mesoscale meteorological model the effects on near-surface variables in an arid metropolitan region. *Journal of Applied Meteorology, 44,* 1281–1297.

Grove, J. M., Cadenasso, M. L., Burch, W. R., Pickett, S. T. A., Schwarz, K., O'Neil-Dunne, J., Wilson, M., Austin, T., & Boone, C. (2006). Data and methods comparing social structure and vegetation structure of urban neighborhoods in Baltimore, Maryland. *Society & Natural Resources, 19,* 117–136.

Harlan, S. L., Brazel, A. J., Prashad, L., Stefanov, W. L., & Larsen, L. (2006). Neighborhood microclimates and vulnerability to heat stress. *Social Science & Medicine, 63,* 2847–2863.

Harlan, S. L., Rex, T., Larsen, L., Hackett, E. J., Kirby, A., Wolf, S., Bolin, B., Nelson, A. L., & Hope, D. (2003). The Phoenix Area Social Survey: Community and environment in a desert metropolis. Central Arizona – Phoenix Long-Term Ecological Research Project Contribution No. 2. Arizona State University.

Harvey, D. (1973). *Social justice and the city.* Baltimore: Johns Hopkins University Press.

Heisler, G. M., & Wang, Y. (2002). Applications of a human thermal comfort model. Fourth Symposium on the Urban Environment, American Meteorological Society, May 20–24, Norfolk, VA. Available at http://www.fs.fed.us/ne/syracuse/Pubs/Downloads/02_GH_ YW_Applications.pdf

Hope, D., Gries, C., Zhu, W. X., Fagan, W. F., Redman, C. L., Grimm, N. B., Nelson, A. L., Martin, C., & Kinzig, A. (2003). Socioeconomic drivers of urban plant diversity. *Proceedings of the National Academy of Sciences of the United States of America, 100,* 8788–8792.

Huete, A. R. (1988). A soil-adjusted vegetation index (SAVI). *Remote Sensing of Environment, 25,* 295–309.

Iceland, J. (2004). Beyond black and white – metropolitan residential segregation in multiethnic America. *Social Science Research, 19,* 325–359.

ICLEI, International Council for Local Environmental Initiatives. (1998). Cities at risk: Assessing the vulnerability of United States cities to climate change. Toronto, Canada. Available at http://iclei.org/co2/car-textonly.htm

IFRC, International Federation of Red Cross. (2003). India: heat wave. *Information Bulletin 1/2003.* Available at http://www.ifrc.org

IPCC, Intergovernmental Panel on Climate Change. (2001). *Climate change 2001: Impacts, adaptations, and vulnerability.* Cambridge, UK: Cambridge University Press.

IPPC, Intergovernmental Panel on Climate Change. (2007). Climate change 2007 summary for policymakers: The physical science basis. Available at http://www.ipcc.ch/ SPM2feb07.pdf

Jenerette, G. D., Harlan, S. L., Brazel, A., Jones, N., Larsen, L., & Stefanov, W. L. (2007). Regional relationships between vegetation, surface temperature, and human settlement in a rapidly urbanizing ecosystem. *Landscape Ecology, 22,* 353–365.

Kalkstein, L. S., & Greene, J. (1997). An evaluation of climate/mortality relationships in large US cities and the possible impacts of a climate change. *Environmental Health Perspectives, 105,* 84–93.

King, D. A. (2004). Climate change: Adapt, mitigate, or ignore. *Science, 9,* 176–177.

Kirkby, J., O'Keefe, P., & Howorth, C. (2001). Introduction: Rethinking environment and development in Africa and Asia. *Land Degradation and Development, 12*, 195–203.

Klinenberg, E. (2002). *Heat wave: A social autopsy of disaster in Chicago.* Chicago: University of Chicago Press.

Kluger, J. (2006). The tipping point. *Time Magazine 167*(April), 34–42.

Kozol, J. (2005). *The shame of the nation: The restoration of apartheid schooling in America.* New York: Crown.

Larsen, J. (2006). Setting the record straight: More than 52,000 Europeans died from heat in summer 2003. The earth policy institute at columbia University. Available at http:/ :www.earth-policy.org/Updates/2006/Update56.html

Larsen, L., & Harlan, S. L. (2006). Desert dreamscapes: Residential landscape preference and behavior. *Landscape and Urban Planning, 78*, 85–100.

Larsen, L., Harlan, S. L., Bolin, B., Hackett, E. J., Hope, D., Kirby, A., Nelson, A., Rex, T., & Wolf, S. (2004). Bonding and bridging: Understanding the relationship between social capital and civic action. *Journal of Planning Education and Research, 24*, 64–77.

Liverman, D., Moran, E. F., Rindfuss, R. R., & Stern, P. (Eds). (1998). *People and pixels: Linking remote sensing and social science.* Washington, DC: National Academy Press.

Logan, J., & Molotch, H. (1987). *Urban fortunes: The urban economy of place.* Berkeley: University of California Press.

Logan, J., Stults, B. J., & Farley, R. (2004). Segregation of minorities in the metropolis: Two decades of change. *Demography, 41*, 1–22.

Lubowski, R., Vesterby, M., Bucholtz, S., Baez, A., & Roberts, M. J. (2006). Major uses of land in the United States, 2002. Economic Information Bulletin No. 14. Economic Research Service, United States, Department of Agriculture.

Luckingham, B. (1989). *Phoenix: The history of a southwestern metropolis.* Tucson: University of Arizona Press.

Martin, C. A. (2001). Landscape water use in Phoenix, Arizona. *Desert Plants, 17*, 26–31.

Martin, C. A., Peterson, K. A., & Stabler, L. B. (2003). Residential landscaping in Phoenix, Arizona, US: Practices and preferences relative to covenants, codes, and restrictions. *Journal of Arborioculture, 29*, 9–16.

Martin, C. A., Warren, P. S., & Kinzig, A. P. (2004). Neighborhood socioeconomic status is a useful predictor of perennial landscape vegetation in residential neighborhoods and embedded small parks of Phoenix, AZ. *Landscape and Urban Planning, 69*, 355–368.

Massey, D. S. (1996). The age of extremes: Concentrated affluence and poverty in the twenty-first century. *Demography, 33*, 395–412.

Massey, D., & Denton, N. (1993). *American apartheid: Segregation and the making of the underclass.* Cambridge, MA: Harvard University Press.

McGeehin, M., & Mirabelli, M. (2001). The potential impacts of climate variability and change on temperature-related morbidity and mortality in the United States. *Environmental Health Perspectives, 109*, 185–189.

Motavalli, J. (Ed.) (2004). *Feeling the heat: Dispatches from the front lines of climate change.* NY: Routledge.

NOAA, National Oceanographic and Atmospheric Administration. (2005). Heat wave: A major summer killer. Available at http://www.nws.noaa.gov/om/brochures/heatwave.pdf

NYCHP, New York Climate and Health Project. (2004). *Assessing potential public health and air quality impacts of changing climate and land use in Metropolitan New York.* New York: The Earth Institute at Columbia University.

Oke, T. R. (1997). Part 4: The changing climatic environments: Urban climates and global environmental change. In: R. D. Thompson & A. Perry (Eds), *Applied climatology principals and practice* (pp. 273–287). London: Routledge.

O'Neill, M. S., Zanobetti, A., & Schwartz, J. (2003). Modifiers of temperature and mortality association in seven US cities. *American Journal of Epidemiology, 157*, 1074–1082.

Park, R. E. (1915). The city: Suggestions for the investigation of human behavior in the city environment. *The American Journal of Sociology, XX*, 577–612.

Patz, J. A., Campbell-Lendrum, D., Holloway, T., & Foley, J. A. (2005). Impact of regional climate change on human health. *Nature, 438*, 310–317.

Pellow, D. N. (2000). Environmental inequality formation: Toward a theory of environmental justice. *American Behavioral Scientist, 43*, 581–601.

Pickett, S. T. A., & Cadenasso, M. L. (1995). Landscape ecology – spatial heterogeneity in ecological systems. *Science, 269*, 331–334.

Pickett, S. T. A., Cadenasso, M. L., Grove, J. M., Nilon, C. H., Pouyat, R. V., Zipperer, W. C., & Costanza, R. (2001). Urban ecological systems: Linking terrestrial ecology, physical, and socioeconomic components of metropolitan areas. *Annual Review of Ecology and Systematics, 32*, 127–157.

Redman, C. L. (1999). *Human impact on ancient environments.* Tucson, AZ: University of Arizona Press.

Redman, C. L., Grove, J. M., & Kuby, L. H. (2004). Integrating social science into the Long-Term Ecological Research (LTER) network: Social dimensions of ecological change and ecological dimensions of social change. *Ecosystems, 7*, 161–171.

Sampson, R. J., Morenoff, J. D., & Gannon-Rowley, T. (2002). Assessing 'neighborhood effects': Social processes and new directions in research. *Annual Review of Sociology, 28*, 443–478.

Sampson, R. J., Raudenbush, S. W., & Earls, F. (1997). Neighborhoods and violent crime: A multilevel study of collective efficacy. *Science, 277*, 918–924.

Schulz, A. J., Williams, D. R., Israel, B. A., & Lempert, L. B. (2003). Racial and spatial relations as fundamental determinants of health in Detroit. *Milbank Quarterly, 93*, 215–221.

Semenza, J. C., Rubin, C. H., Falter, K. H., Selanikio, J. D., Flanders, W. D., Howe, H. L., & Wilhelm, J. L. (1996). Heat-related deaths during the July 1995 heat wave in Chicago. *The New England Journal of Medicine, 335*, 84–90.

Sethi, R., & Somanathan, R. (2004). Inequality and segregation. *Journal of Political Economy, 12*, 1139–1147.

Smoyer, K. E., Rainham, D. G., & Hewko, J. N. (2000). Heat-related stress mortality in five cities in southern Ontario: 1980–1996. *Institutional Journal of Meteorology, 44*, 190–197.

Speth, J. G. (2004). *Red sky at morning: America and the crisis of the global environment.* New Haven: Yale.

Stefanov, W. L., Prashad, L., Eisinger, C., Brazel, A., & Harlan, S. L. (2004). Investigation of human modifications of landscape and climate in the Phoenix Arizona metropolitan area using MASTER data. *The International Archives of the Photogrammetry, Remote Sensing, and Spatial Information Sciences, 35*(B7), 1339–1347.

Syme, G. J., Shao, Q., Murni, P., & Campbell, E. (2004). Predicting and understanding home garden use. *Landscape and Urban Planning, 8*, 121–128.

Veblen, T. (1899). *The theory of the leisure class.* New York: Macmillan.

Voogt, J. A. (2002). Urban heat island. In: I. Douglas (Ed.), Causes and consequences of global environmental change (Vol. 3, pp. 660–666). *Encyclopedia of global environmental change.* Chichester, UK: Wiley.

Watson, R. T., & the Core Writing Team. (2002). *Climate change 2001: Synthesis of the third assessment report of the intergovernmental panel on climate change.* Cambridge, UK: Cambridge University Press.

WHO, World Health Organization Task Group. (1990). Potential health effects of climate change. World Health Organization, Geneva, Switzerland. Available at http://www. ciesin.org/docs/001-007/001-007.html

Wilson, W. J. (1987). *The truly disadvantaged: The inner city, the underclass, and public policy.* Chicago: University of Chicago Press.

Yabiku, S., Casagrande, D., & Farley-Metzger, E. (in press). Preferences for landscape choice in a southwestern desert city. *Environment and Behavior.*

EQUITY, VULNERABILITY AND RESILIENCE IN SOCIAL–ECOLOGICAL SYSTEMS: A CONTEMPORARY EXAMPLE FROM THE RUSSIAN ARCTIC

Bruce C. Forbes

ABSTRACT

Environmental and social problems are tightly linked in coupled social–ecological systems in the Arctic. This chapter will discuss the importance of equity as a factor in the adaptive capacity of a region undergoing relatively rapid climate change and simultaneous land-use change (petroleum development) in the northwest Russian Arctic. Relative to North America, attempts to implement some kind of economic or legal equity with regard to massive industrial development are token at best. Unfortunately, in the current situation, legal rights to land and resources are neither likely to materialize nor, even if they did, to facilitate adaptive capacity on the part of Nenets herders. As such, herders lack power over important decisions pertaining to the manner in which development proceeds on their traditional territories.

Equity and the Environment
Research in Social Problems and Public Policy, Volume 15, 203–236
Copyright © 2008 by Elsevier Ltd.
ISSN: 0196-1152/doi:10.1016/S0196-1152(07)15006-7

Russia's northern lands have been developed along starkly different lines than those of Europe and North America. Yet the limited literature of resilience in northern social–ecological systems is derived almost exclusively from North American experiences with co-management. Recent work on the Yamal Peninsula indicates that even with a sustained commitment to active engagement, only incremental change is expected. Western-style legislative campaigns and overnight blanket solutions are far less likely to bear fruit and may, in fact, be counterproductive. The prescriptive approaches from four different analyses of the Yamal situation are compared, with special attention devoted to their respective assessments of resilience. Fortunately, the retention of youth within the nomadic population of tundra Nenets appears to be high, providing a positive indicator of overall resilience in this particular social–ecological system.

BACKGROUND AND INTRODUCTION

By definition, environmental and social problems are strongly linked in so-called "social–ecological systems" or SESs (Folke et al., 2002; Folke, Berkes, & Colding, 2003). Some have argued that this is especially the case in northern high-latitude regions where, at the dawn of the 21st century, many indigenous and non-indigenous people are still dependent to a greater or lesser extent on marine and terrestrial wildlife for some combination of food, clothing, shelter and spiritual fulfillment, as they have been for millennia (Berkes, 1998; Berkes & Jolly, 2001; AHDR, 2004; Chapin et al., 2004, 2006a, 2006b). This is particularly the case in northern Russia, where certain populations of hunters, herders, fishers and gatherers continue to live their lives largely "on the land," migrating almost constantly within and among the tundra and taiga, much as they have for at least 1,000 years. These resource relationships, generally but not exclusively with animals, are so strong as to serve, along with technological change, as markers for cultural differentiation in space and time (Helm, 1981; Damas, 1984; Krupnik, 1993). Passing knowledge of these cultures, coupled with images of spectacular landscapes and seascapes, have helped to create a popular vision of the Arctic as a bleak wilderness sparsely settled by primitive cultures. This view is misleading at best, patronizing and counterproductive to building resilience at worst (Forbes, 2005a). As active hunters, herders, fishers and gatherers, obligate rural groups around the Arctic are already facing special challenges from climate change and globalization processes that more urban and suburban segments of the population generally are not likely to ever experience in the same manner (Smit & Pilifosova, 2001; Nuttall et al., 2005). Although northern communities keenly observe the waxing and waning of populations of both wild and domesticated animals and plants, it is only in the past few years that the Arctic has entered into the public consciousness as a bellwether of changes that may affect, or are already affecting, the rest of the globe (ACIA, 2005).

As in other parts of the world, attempts to "manage" and "conserve" the natural resources of the Arctic have met with decidedly mixed success (Forbes, 2005b). The Society for Conservation Biology has recently acknowledged that, nearly two decades after its founding, more species and ecosystems than ever are at risk globally, despite undeniable progress in our understanding of ecological patterns and processes at several scales. In 2002, the Society began a frank internal discussion over the reasons for this perceived failure and concluded that it is not because of bad science, although there is always plenty of that to go around. Rather, the disconnect between our ecological knowledge and conservation success has derived in large part from a general inability among natural scientists to accept that social factors are often the primary determinants of success or failure (Mascia et al., 2003). For natural scientists undertaking ostensibly policy-relevant research, it has been difficult over the past 2–3 decades to accept that no matter how many new regulations, migration corridors and protected areas are established – each one an ostensible "success story" in its own right – ecosystems continue to degrade globally and large numbers of species edge toward and on into extinction.

It is obvious to even the casual observer that the Arctic is not characterized by the levels of biodiversity found in more temperate and tropical regions (Olson & Dinerstein, 1998). The Arctic is, however, home to large and widespread populations of wildlife such as caribou/reindeer and marine mammals that, in turn, have supported humans for thousands of years (Krupnik, 1993; Forbes & Kofinas, 2000). The region is therefore extremely rich in cultural diversity (Damas, 1984; Vakhtin, 1992; AHDR, 2004). Nonetheless, as the general public and at least some politicians have come to recognize anthropogenic global change as an issue, popular attention has predictably focused on the potential loss of "keystone" species, e.g. polar bears, caribou and whales, rather than the cultures that have developed in intimate association with these animals over centuries and millennia. In the global hoopla surrounding the recent release of the Arctic Climate Impact Assessment (ACIA, 2005), it appeared at times like more attention was being paid to the rapidly disappearing tundra, glaciers and permafrost than to the equally serious threats to arctic indigenous cultures.

After the overwhelming and largely sympathetic attention afforded the concerns raised by ACIA, at least one U.S. congressman and the front page of *The Wall Street Journal* were compelled to question skeptically how truly variable had arctic climate been in the past (Regalado, 2005). Suddenly, the Arctic was everywhere in the news, for a few weeks or months anyway, and it seemed to be quite vulnerable. Images of open water at the North Pole in summer 2004 were followed by news of late freezing sea ice in November 2004 and then early melting sea ice in spring 2005. These, in turn, were followed by the indelible images of hurricane Katrina in August 2005. Rightly or wrongly, the media and a great many people began to try to connect the dots based on this short burst of occasionally misguided media attention to long-term, complex phenomena. The problems facing the polar bears, the Inuit and the permafrost were no longer limited to the "remote" Arctic. Global change, in particular climate change, was something that could threaten whole cities far from the Arctic. Suddenly, the entire planet seemed vulnerable.

In the wake of Katrina, experts of different stripes proclaimed that they had warned for years that the levees of the Mississippi River delta were highly susceptible to collapse in a severe storm. At the same time, politicians declared confidently that the destroyed urban areas of the Gulf of Mexico coast could and would rebuild, the people would return and that the entire region would recover. Without using the precise words, the concepts of "vulnerability" and "resilience" lay at the core of these arguments. The sad truth is that levels of vulnerability and resilience in cities like New Orleans were driven in large part by socio-economic and racial equity or, more precisely, the lack thereof. Despite the strenuous denials of politicians, the disparity in terms of storm impact and response across different neighborhoods was obvious to all.

These examples are presented to make the following points. Public acceptance, if not understanding, of the Arctic, as a region critically important to the global climate system, is probably at an all-time high. Equally important, if not generally acknowledged, is the specter of degradation in combined social–ecological systems across significant portions of the Arctic, particularly where rapid climate and land-use change interact. The track record of ecosystem conservation across the globe is admittedly poor and getting worse, in large part because human actions are not properly taken into account when attempting to "manage" ecosystems. These concepts will be explored using the example of a nomadic pastoralist group in the Russian Arctic, the tundra Nenets of Yamal (Fig. 1). For them, inequality is manifest in the form of limited ability to leverage meaningful

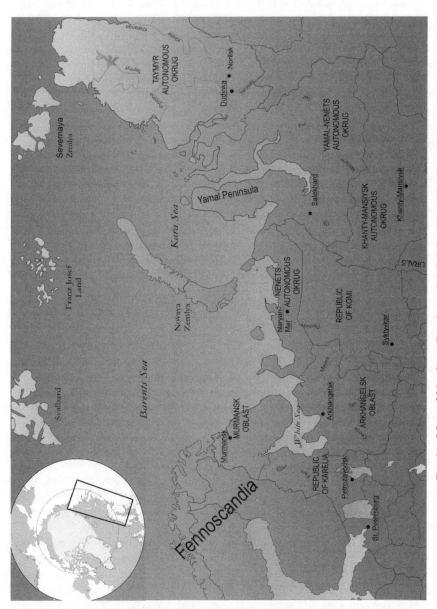

Fig. 1. Map of Northern Russia and Adjoining Northern Europe.

consultation and powers of decision-making that would mitigate the negative impacts of development on their traditional territories.

YAMAL NENETS REINDEER HERDERS OF NORTHWEST SIBERIA

All around the circumpolar Arctic, indigenous peoples tend to see themselves as integral ecosystem components in the areas where they reside (AHDR, 2004; Nuttall et al., 2005). In particular, nomadic reindeer herders of the Yamal-Nenets Autonomous Okrug[1] who live north of the arctic treeline immediately east and north from Russia's Ural Mountains, exemplify this spirit of being, quite literally, close to the land (Stammler, 2005). The Yamal Nenets are among the few remaining truly nomadic pastoralists. There is no need to romanticize this special relationship. The simple fact is that migratory herders spend most of their lives on the tundra and in close contact with their reindeer, apart from time spent as children in school and, for young men, engaged in compulsory military service. They also spend significant amounts of time fishing, hunting and gathering. This time on the land year after year allows for, or rather necessitates, the development and maintenance of highly complex social and ecological skills and forms of knowledge that cannot be learned in any classroom. Nor can such "ways of knowing" about the land be retained in perpetuity if and when younger generations migrate to towns and cities. A large literature has developed in recent decades around the concepts of "traditional," "local" or "practical" ecological knowledge, but this will not be reviewed here (see, however, Berkes, 1998, 1999; Usher, 2000; Kendrick, 2003a, 2003b, 2003c; Huntington, Callaghan, Fox, & Krupnik, 2004; Kitti, Gunslay, & Forbes, 2006). Suffice to say that the tundra Nenets' situation contrasts with that in arctic North America, where virtually all indigenous peoples were relocated into fixed settlements by the late 1950s and early 1960s for purposes of sovereignty, education, religious indoctrination, law enforcement and, not least, the fur trade.

On the one hand, arctic nomadism exemplifies resilience. The ongoing harvest of caribou/reindeer (*Rangifer tarandus*) and marine mammals has allowed different groups over millennia to occupy, extensively and persistently, a huge and climatically diverse and dynamic region that seems to most outsiders to be remote and forbidding (Krupnik, 1993; Forbes & Kofinas, 2000). On the other, certain arctic indigenous peoples are generally considered to be extremely vulnerable to the newly synergistic forces of rapid climate and land-use change (AHDR, 2004; ACIA, 2005). Entering the 21st century, the Yamal Nenets continue to migrate with the reindeer much as they have for countless generations, cued by – among other things – the cyclic greening and senescing of tundra vegetation, the melting and re-freezing of ancient rivers and the appearance and disappearance of biting insects (Stammler, 2005). Their homeland is effectively sandwiched between the latitudinal treeline to the south and the Arctic Ocean to the north. Fully cognizant of the modern world at their doorstep, they consciously choose to keep it at arm's length, taking what they need and, with increasing difficulty, attempting to keep out what they do not (Stammler, 2002). Their strong sense of independence and expert abilities as herders has served them well through successive Tsarist and Soviet regimes (Golovnev & Osherenko, 1999). Other factors include their nomadic lifestyle, requiring daily use of a fund of traditional knowledge, economic autonomy and a minimalist ethic. Further, Golovnev and Osherenko (1999) point out that these together "generate flexibility rather than rigidity," and that this flexibility is reflected in the Nenets' pattern of leadership and in gender roles. In many ways, they have demonstrated marked adaptive capacity and resilience in the face of massive systemic shocks, such as the onslaught of collectivization and the subsequent collapse of the Soviet Union. In addition, their success is partly attributed to the fact that Soviet pressure for them to submit to a sedentary life was not as great as for peoples in other regions (Stammler, 2005). In any case, active adaptation is essential in the current highly dynamic situation (Klokov, 2000).

Widespread industrial development was not present in the Arctic prior to World War II. Possible exceptions include the mining of minerals on Kola Peninsula (Murmansk region) and the complex of coal mines and cement factories in and around Vorkuta (Komi Republic), both extensively developed with forced labor under Stalin beginning in the 1930s. Oil and gas development is therefore a regional newcomer that threatens Nenets in ways unlike earlier encroachments into their territory. In a typical contemporary Russian oil or gas field, the extent of ecosystem degradation is often an order of magnitude greater than in a comparable Canadian or Alaskan field. Yet such impact pales next to the upheaval that all Russians have experienced since the collapse of the Soviet Union. Equity is a key issue facing the Nenets as they ponder their own vulnerability in the face of extensive ecosystem reorganization and, most likely, degradation, over which they so far have little or no control (Chance & Andreeva, 1995).

Aspects of incremental climate change are intermittently evident, such as warmer winters and early springs in recent years and an overall warming of ~3°C since the 1970s (ACIA, 2005), but are necessarily pushed to the background as they confront the unrelenting challenges of weekly and monthly surprises associated with accelerating petroleum development.

The territory of the Yamal Nenets overlaps directly with some of the largest untapped natural gas and gas condensate deposits in the world. Supergiant gas fields on the Yamal Peninsula have been intensively explored since the late 1970s and are currently gearing up for full-scale production, primarily under the state-run enterprise Gazprom and its respective subsidiaries. Parallel efforts are underway to exploit huge oil deposits in the neighboring Nenets Autonomous Okrug (Stammler & Forbes, 2006). The Russian state has set itself ambitious goals for increasing oil and gas production and export within the next 5–10 years. Distribution of gas from Yamal to Russian, European and eventually North American markets is expected to take place via a combination of overland pipelines and tanker shipping through the northern sea route. The land and near-shore coast are underlain by frozen ground, or continuous permafrost, often some hundreds of meters deep. Much of the land is characterized by so-called ice-rich permafrost (ACIA, 2005). This and the extreme cold in winter seriously complicates numerous aspects of engineering and infrastructure development and may well impinge upon the timeline set for getting into full production mode (Talley, 2006). In the short term, however, development proceeds apace and the Nenets are working hard to engage the state, its various enterprises and proxy intermediaries head on in an attempt to preserve their way of life. For those unfamiliar with modern life in northern Russia, the strong contrast with western social–ecological systems and legal norms is perhaps informative.

RESILIENCE: EXAMPLES FROM CO-MANAGEMENT IN SOCIAL–ECOLOGICAL SYSTEMS

To assess the issues of resilience in a region of contemporary arctic Russia, we must first briefly review the relevant literature, as well as the history of interactions between northern peoples and southern institutions. The literature of resilience and vulnerability is vast and growing. In a recent review of nearly 3,000 papers published over a 38-year period, Janssen, Schoon, Ke, and Börner (2006) reported a rapid proliferation in publications between 1995 and 2005. They observed that "the resilience knowledge domain has a background in ecology and mathematics with a focus on theoretical models, while the vulnerability and adaptation knowledge domains have a background in geography and natural hazards research with a focus on case studies and climate change research." They also detected an increasing integration of the different knowledge domains, perhaps reflective of the rise in multidisciplinary approaches to research.

In accordance with the findings of Janssen et al. (2006), most of the empirical publications on resilience in the Arctic come from the general domain of ecology, in particular disturbance ecology and environmental change (e.g. Felix, Raynolds, Jorgenson, & DuBois, 1992; Reynolds & Tenhunen, 1996; Wookey & Robinson, 1997; Chapin et al., 2003). These works emphasize the absence of equilibrium or, rather, the ability of systems to switch between different states. As such they adhere to what Holling and Gunderson (2002) term "ecological resilience." At any rate, while rich and undoubtedly relevant, the resilience and vulnerability literature is relatively shallow when it comes to examples from combined social–ecological systems in arctic or subarctic regions. Empirical studies are few, treating resilience rather than vulnerability, and most tend to be derived from North America (e.g. Berkes, 1998; Berkes & Jolly, 2001; Kendrick, 2003a). These examples overlap considerably geographically and socio-culturally with the literature on co-management (Kruse, Klein, Braund, Moorehead, & Simeon, 1998; Klein, Moorehead, Kruse, & Braund, 1999; Usher, 2000; Kendrick, 2002; Peters, 2003; Berkes, Bankes, Marschke, Armitage, & Clark, 2005). Co-management is defined as a shared decision-making process, formal or informal, between a government authority and a user group, for managing a species of fish and wildlife, or other resources (Caulfield, 2000). This shared process is a form of equity in decision-making that is normally held by the state. There are many different types of co-management, as reviewed by Berkes (in press).

Taken as a whole, there is a strong overlap between the literature on co-management, resilience and that of "local" or "traditional ecological knowledge" (i.e. TEK) (Treseder, 1999; Huntington, 2000; Berkes & Folke, 2002; Kendrick, 2002, 2003a, 2003c; Peters, 2003; Moller, Berkes, Lyver, & Kislalioglu, 2004). This is because a cornerstone of co-management is the integration of western science and traditional knowledge via the translation of indigenous life experiences into forms compatible with state wildlife management (Nadasdy, 2003). We must therefore accept a strong North American bias in these respective literature domains. While the cited examples of resilience in social–ecological systems are drawn from a diverse set of conditions across arctic and subarctic North America (Berkes, 1998; Berkes & Jolly, 2001; Kendrick, 2003a), it seems that none of them are truly relevant to contemporary Europe and Russia. One published study does concern adaptive co-management in Europe (Olsson, Folke, & Berkes, 2004), but it is from southern Sweden and does not confront a suite of factors similar to those found in truly arctic social–ecological systems. Others concern North European fisheries and reindeer herding, but so far, only Norwegian fisheries approach something like co-management as it is practiced in North America (Jentoft, 1998).

There are several difficulties with transferring models of co-management and resilience across the Arctic. To begin with, while "virtually every co-management case study encountered in the literature is a success story," co-management has its own problems (Nadasdy, 2003, in press). These stem from the political and other institutional obstacles to truly integrating indigenous knowledge into the dominant "scientific" paradigm of wildlife management embraced by modern states. Those studies that do treat with caution the lessons of co-management, and related research on traditional ecological knowledge, all emphasize that "trust" is the critical factor in finding any kind of common ground between indigenous "users" and scientists (Ferguson & Messier, 1997; Kendrick, 2003a, 2003b; Nadasdy, 2003; Olsson et al., 2004). Without trust, there can be no real equity in shared decision making. The danger is that despite the rhetoric of local empowerment that generally accompanies such processes, they often actually serve to perpetuate colonial-style relations by concentrating power in administrative centers, rather than in the hands of local/aboriginal people (Nadasdy, 2007). Once a co-management regime is in place, defining "success" becomes an additional problem (Nadasdy, 2003).

Another factor is that North American co-management regimes have relied heavily on the institution of "elders" as holders of knowledge, who therefore tend to be chosen to represent community-level understanding and concerns as they pertain to wildlife and other issues (e.g. protected areas). In Russia, there is no council of indigenous elders, and Nenets who have visited North America found the practice amusing. In their eyes, elders would not know how to talk to state representatives or companies. Quite on the contrary, crucial partners to implement coexistence in Russia are either Soviet-educated indigenous intelligentsia (most of whom are not really elders), or, increasingly, young, dynamic emergent leaders. While this hints at significant inequalities within Nenets society, in fact herders seem reasonably satisfied with their chosen mode of representation by these two main groups. During Soviet times, the influence of pensioners was nominally diminished as they were deprived of official decision-making power in administrative matters. Nonetheless, elders' knowledge retains its functional importance, since people still consult elders in relation to life on the tundra. Stammler (2005) argues that in many cases, Soviet rule abandoned kinship relations and a Nenets system of leadership in the tundra only superficially, while the role of respected leaders and the importance of kinship relations continued to be important throughout the Soviet period. However, respect in Nenets society was not acquired by mere age, but more by particular qualities of a character.

At its core, co-management is based on participatory approaches to policy-directed research that involves local users, indigenous or not. There is a long history of co-management in North America (Treseder, 1999). It is only in recent years that discussion of truly participatory approaches to northern resource management has taken place in the terrestrial ecosystems of subarctic Europe, in each case involving reindeer management (Karlstad, 1998; Sandström et al., 2003; Forbes et al., 2006). Experiments in participatory research have encountered varying degrees of resistance from European states (e.g. Sweden, Finland). This is in large part because natural resource management has historically been the exclusive domain of natural scientists, whose research is conducted at the behest of the relevant ministries (forestry, agriculture, environment) to advance their respective agendas, aspects of which are considered strongly paternalistic by Sámi reindeer herders (Forbes, 2006; Müller-Wille et al., 2006). Russia, as Stammler (2005) points out, also has its own history in which the "outside world" has structured its interactions in a paternalistic manner with what he refers to as the "tundra sphere." Given these differing histories, and the limitations of the existing literature, we need to be cautious in our assumptions about the applicability of earlier equity, vulnerability and resilience analyses to new situations. The purpose of introducing the case study of the Yamal Nenets here is to provide some geographic, cultural and historical contrast to the North American institutional interactions that prevail in the literature.

CONTEMPORARY REINDEER MANAGEMENT: EQUITY IN DECISION-MAKING

Rangifer tarandus L. *sensu lato* is the species of the deer family Cervidae that includes all wild and semi-domestic forms of caribou and reindeer in the world (Harrington, 2005; Forbes, 2005c). Human–*Rangifer* systems are circumpolar in the northern hemisphere and encompass a variety of coupled social–ecological systems (Forbes & Kofinas, 2000). The diversity of these systems is striking. They are dominated by hunting cultures in North America, whereas in Europe and Russia there exists a mix of herding and hunting cultures. Importantly, there have been identified continua within, and occasional transitions among, these general designations over time (Ingold, 1980; Krupnik, 1993; Baskin, 2000; Müller-Wille et al., 2006). Furthermore, within Russia, this categorization is institutionalized, in that hunting is mostly considered a less prestigious occupation than herding (Stammler, 2005). In either context, *Rangifer* is an extremely complex animal ecologically and behaviorally. Nonetheless, its circumpolar distribution and interaction with virtually all arctic and subarctic cultures makes it tempting to draw comparisons between modern day management regimes from different regions.

Evidence from a large research project in northern Fennoscandia[2] clearly illustrates the contrast between recent North American and European experiences with regard to participatory institutions for resource management (Forbes, 2006; Hukkinen et al., 2006). The project, RENMAN, was an experiment in participatory action research in northernmost Europe, where indigenous Sámi reindeer herders and their respective states (Norway, Sweden, Finland) have historically confronted similar dilemmas of resource equity for all citizens versus resource protectionism for indigenous minorities (Jernsletten & Beach, 2006). Compared to their counterparts under co-management regimes in North America, the Sámi have little equity concerning individual or collective decisions concerning short- and long-term management of their herds.

Along with other societies in northern Russia, the Yamal Nenets experienced major shocks in the last century, which affected their ability to maintain their traditional way of life. The most obvious is collectivization, which began in the late 1920s, when reindeer herding became a branch of agriculture. As happened with Sámi reindeer herding after World War II, the aim was to turn reindeer into a commodity suitable for the market and to provide state institutions with meat (Stammler, 2005; Hukkinen et al., 2006; Kitti et al., 2006). Both Russia and European states attempted to wrest control of the seasonal change of pastures and implemented a management model based on "carrying capacity," although honest biologists admit how difficult it is to evaluate actual carrying capacity (Helle & Kojola, 2006; see also Mysterud, 2006, van der Wal, 2006). The difference was that Sámi were to be engaged in herding as an enterprise, whereas in Russia, all property was confiscated and reindeer ownership transferred to the state.

While accepting that significant ruptures took place, Stammler argues that, in the end, things changed relatively little, using first the examples of reindeer pasture migration routes. Golovnev and Osherenko (1999) suggest that migration routes were radically altered. This observation derives from the conversion from kolkhozy to sovkhozy in the 1960s, which cut off southern migratory area from the northernmost Yamal sovkhoz. However, based on extensive fieldwork in the Yarsalinski sovkhoz, which comprises the largest and longest of the region's sovkhoz territories, Stammler (2005, p. 131) concludes "that the general pattern of migration did not change and that migration is actually one of the main examples of continuity from pre- to post-Soviet times." Other aspects of life that he notes remained unchanged include de facto private ownership of animals throughout the entire Soviet period, the use of the *chum* (a traditional form of tent or teepee) as mobile housing during migration, and the use of reindeer for transport rather than motor vehicles. He attributes this to the Nenets' ability to adapt the Soviet system as needed and use it to their advantage, as well as to the fact that pressures from Soviet authorities to submit to a sedentary life were apparently not as great for Nenets as for other indigenous peoples. Golovnev and Osherenko (1999) hold that Nenets indeed were pressured to give up land and resources, and responded by remarkable rebellions in the 1930s incorporating ritualistic practices.

Concerning animal ownership, the rate of private to sovkhoz (publicly owned) reindeer shifted from being 30 percent held as private stock in the Soviet period to 70 percent private in the post-Soviet 1990s. Retention of the sovkhoz system was a conscious decision on the part of Nenets who held positions of leadership in the Yarsalinsky sovkhoz. The leaders understood that without state subsidies through the sovkhoz, the herders would not have access to cash needed to buy and deliver basic supplies to tundra camps. The sovkhoz system allowed continuous payments to herders and "tent workers" without which the migratory herders would have had an increasingly hard existence. The success Nenets have had to date is in significant measure due to the insight of Nenets' leaders. In many or most other areas, people completely dismantled the state farm system (Golovnev & Osherenko, 1999).

Another example comes from the institutional restructuring of collective herds (meaning Soviet state farms or *sovkhozi*) that took place upon the collapse of the Soviet Union. Yamal herders refer to this time during the 1990s as "the decade of chaos" (Stammler, 2005). This period was characterized by a confusing struggle between privatization and state ownership of agricultural enterprises in which some collective herds were nominally liquidated and others survived. Sovkhoz herds did decline in the 1990s but since 2000 have increased. In many respects, the sovkhoz structure was retained. Stammler (2005) cites the institutional change of the sovkhoz herds as evidence of continuity on Yamal, in spite of the undoubted transformations that took place in the post-Soviet restructuring.

If one accepts that the Yamal Nenets social–ecological system was subject to significant shocks (Golovnev & Osherenko, 1999; Stammler, 2005), then resilience is demonstrated by sustaining those attributes that are important to society in the face of change. According to Chapin et al. (2006a), adaptation means developing new socio-ecological configurations that function effectively under new conditions. Stammler (2005) echoes this with regard to the Yamal Nenets. As described above, he analyzes their behavioral and historical adjustments to their changing natural, social and economic environment and concludes that their adaptive capacity or "flexibility" has proven to be high in many historical crisis situations. However, he also stresses that this capacity is not unlimited and is indeed threatened by ongoing oil and gas development, which "holds dangers for the future." This is because although the pre-Soviet, Soviet and post-Soviet regimes all needed the cooperation of the Nenets to manage the region for its resources (furs, meat, fish), their partnership in oil and gas extraction is not needed. Indeed, their presence and their hopes for an equitable mutual coexistence have come to be seen by some as problematic. This is despite the fact that in Soviet industrialization ideology, such coexistence was implied and desired, albeit on unequal terms. The past paternal tendencies cited by Stammler are now manifest by state efforts, on the one hand, unilaterally to withdraw lands essential to maintaining traditional migrations, and on the other to facilitate depopulation of the tundra by subsidizing herders' relocation into villages and towns.

CHALLENGES TO MAINTAINING ADAPTIVE CAPACITY

In my own experience in northwest Russia over the past 16 years, climate rarely comes up among Nenets as an important agent of change. According to Nenets themselves, petroleum development presents a more urgent and immediate threat (Khorolya, 2002). The existing and future threats must therefore be placed in the context of the immediate situation, using historical developments as guide, while understanding that the current climate warming begun in the 1960s and 1970s may continue to accelerate and so pose new challenges (ACIA, 2005). As mentioned earlier, in this highly dynamic situation, active rather than passive adaptation is essential (Klokov, 2000).

The push for oil and gas development is a common thread connecting northern Alaska, northwest Canada, and the Russian North. Indigenous identity is tied in part to reliance on the land and sea, and knowledge of how to live there over many generations with or without large subsidies or interference from outside the region (Nelson, 1969; Berger, 1985; Wenzel, 1991). As in Russia, the proven and potential natural gas reserves of arctic Alaska and Canada occur mainly on lands inhabited by indigenous peoples referred to, respectively, as Alaska Natives and First Nations. The difference is that at the time oil was discovered by outsiders in the late 1960s and early 1970s, the political clout of the diverse indigenous groups and their supporters in Washington and Ottawa was already sufficient to instigate a spate of major land claims that were seen as both pioneering and generous as the time. In retrospect, such agreements were not without serious flaws (cf. Berger, 1985; Flanders, 1989). In the case of Alaska, development was allowed to proceed quickly via the Alaska Native Claims Settlement Act (ANCSA) of 1971, which transferred title to 18 million hectares of land and $962.5 million to Alaska Natives. As for the Mackenzie Valley pipeline, its development was shelved in the 1970s until land claims could be negotiated. By the late 1990s, most of the First Nations groups that once opposed drilling on legal, environmental and cultural grounds had abandoned their protests and now strongly favor exploiting their region's energy resources. It is a measure of their hard-won political maturity that these North American groups have attained such a high level of legal and economic strength when it comes to development rights.

With regard to northern energy development, both the Alaskan and Canadian situations have been intimately tied to a degree of political and legal parity in negotiations more or less absent in Russia. To be sure, this is in part due to the timing of the settlements. No one should be under any illusions that if arctic oil fields in the West had been developed during or prior to World War II, the indigenous groups in North America would have received such equitable treatment. Yet the fact remains that at present Russia contrasts in almost every dimension of "human development" (social, cultural, economic, political, legal) with other circumpolar rim countries (AHDR, 2004). Perhaps most importantly for this discussion, the legal rights of arctic indigenous peoples have developed along starkly different lines around the circumpolar North. These range from self-governance in Greenland and Nunavut, to the aforementioned land claims in western North America, to the more limited power held by indigenous northern Russians (Bankes, 2004). In Fennoscandia, the national Sámi parliaments that exist are able to suggest but not enact legislation. Nonetheless, certain hunting, fishing and herding rights are attached to legal definitions of "being Sámi" (Karppi & Eriksson, 2002; Müller-Wille et al., 2006).

Despite these rather sharp regional differences, broadly speaking there are some historical parallels in the perceptions of, and behavior toward, the Arctic. In general, governments have traditionally treated the North as a hinterland for supplying resources to southern populations (Armstrong, Rogers, & Rowley, 1978; Young & Osherenko, 1992). Many, but not all, countries have also promoted the view of the Arctic as a remote and fragile "wilderness" (Forbes, 2005a). The latter is a western concept at odds with Arctic indigenous peoples' perceptions of their own environment. In truth, Nenets see themselves as integral components of the ecosystems they inhabit, which they consider to be bountiful. Conversely, companies and not a few politicians have tended to adopt a view of the Arctic as a barren wasteland. Alongside the concept of wilderness comes the view among many conservationists that the best way to "protect" the Arctic is to establish protected areas.

These perspectives represent various aspects of thinking by outsiders, displaying a rather circumscribed view of humans and nature, with each having its own elements of paternalism. Since the breakup of the Soviet Union, the concept of protected areas and the marketing of "wilderness" tourism have also gained currency in Russia (Forbes, Monz, & Tolvanen, 2004a; Forbes, Fresco, Shvidenko, Danell, & Chapin, 2004b; O'Carroll & Elliot, 2005). According to World Wide Fund for Nature (WWF), their quest to create a circumpolar network of protected areas is making the fastest progress at present in the Russian Arctic. In the last seven years, the territory designated as "protected" has more than doubled, bringing the total reserve land area to about 35 million hectares today – an area the size of Norway (WWF, 2006). However, it has been argued that the establishment of protected areas is often "in conflict with the interests of indigenous peoples and may sometimes adversely affect their prosperity and welfare" (Bolshakov & Klokov, 2000). As with petroleum development, the nascent legislative sphere concerning indigenous peoples' rights in relation to arctic ecosystems presents both dangers and opportunities. Officially, the state does recognize that each sovkhoz has "unlimited use rights" for herding, which also encompasses rights for hunting, fishing and gathering. Yet even the presence of significant numbers of migratory Nenets in the tundra does not necessarily serve to limit the pace and extent of development (Stammler, 2005). It merely adds another step of complication in pushing development forward. This requires negotiating the transfer of "agricultural land" (e.g. pastures) of an authorized long-term user (e.g. sovkhoz) into industrial land (e.g. for extraction) to be used by a new limited-term user (e.g. an oil company). This often involves much paperwork. But if the development project is important for somebody with power, these things can proceed surprisingly quickly (F. Stammler, personal communication).

PRESCRIPTIVE APPROACHES

Although they generally do not use the language of vulnerability and resilience, authors who have analyzed the Yamal Nenets' particular situation have recommended different approaches to the dilemmas presented by ongoing oil and gas activities (Chance & Andreeva, 1995; Golovnev & Osherenko, 1999; Zen'ko, 2004; Stammler, 2005). Equity in some form or another is at least considered by Chance and Andreeva (1995) and (Zen'ko, 2004). Given the aforementioned variations in the development of contemporary legal and political regimes, sensitivity to the special Russian context is essential.

To begin with, Golovnev and Osherenko (1999) outline three models "for restoring rights and decision-making authority" to the Nenets, drawn from three broad categories: tribal governments, public governments where the indigenous population constitutes a majority (e.g. Home Rule in Greenland) and co-management. They readily admit that none of these three are perfect solutions and all of them "risk undermining indigenous leadership and destroying existing modes of decision-making."

Using some different historical examples than Stammler (2005), Golovnev and Osherenko (1999) make the same basic point concerning the relatively strong adaptive capacity of Yamal tundra Nenets. In the final paragraph of their analysis they stress that, "Nenets nomads who travel lightly across the tundra have the internal quality of flexibility to adapt to new conditions." They further suggest that "outsiders must allow them the space for adaptation and the opportunity to retain control over their lives." To an uninformed westerner, the latter appears as a rather passive outlook, in which the assumption seems to be that the respective national, regional and sub-regional governments, along with the thousands of newcomers arriving in the region each year, are capable of acting as essentially benign forces. Far from benign, the lead company active on Yamal Peninsula – Nadymgazprom – actively exploits weaknesses in the unresolved nature of enforcement, as well as the lack of monitoring by law enforcement agencies and federal regulatory services (Zen'ko, 2004). Typical western alternative responses might range from vocal environmental and legal activism to more quiet rebellion among significant numbers of concerned shareholders. The former almost certainly represent the wrong track and are instead likely to backfire. The latter is impossible given that Gazprom and its various daughter enterprises are functionally state held.

As was mentioned earlier, Russia's indigenous peoples do not have the same level of political clout as those in western countries. They reside in radically different systems, with their own histories and cultures. In many ways they are simply not comparable. As with the political systems, so go the legal and corporate structures and environmental responsibilities. To put this into perspective, non-Russians must appreciate that functionally state held entities like Gazprom are not beholden to international shareholders. Although there has been some improvement in recent years, these powerful monopolies to date have expressed little concern with regard to their environmental image. Multinational British Petroleum, on the other hand, has engaged in an extended, costly and strategic campaign to re-brand itself as an environmental champion "beyond petroleum" (Whiteman, Forbes, Niemelä, & Chapin, 2004). Thus its image was hit hard by the August 2006 shutdown of its Prudhoe Bay operations due to pipeline leakage (Hoyos, Bream, Harvey, & McNulty, 2006; Mufson, 2006).

Yet even western partnership does not necessarily ensure the implementation of best practices on the ground in Russia. For mutual coexistence, the real onus is on the local and regional actors of state and joint venture enterprises to accommodate indigenous peoples, who themselves posses little real power in the present system. It is very hard work on all sides and nomadic reindeer herders understand and accept that it is impossible for development to have no impact. The Alaska pipeline has shown clearly that even with stringent standards dutifully enforced, some direct and indirect or cumulative impacts are inevitable (NRC, 2003; Mufson, 2006). Yet people on Yamal know from their past experience with Amoco in the 1990s that it is possible to have meaningful consultations and to implement research that addresses local concerns, such as monitoring conditions of reindeer pastures, rehabilitation of disturbed areas and cultural heritage preservation

(Martens, 1995; Martens, Magomedova, & Morozova, 1996; Fedorova, 1998).

In separate publications, Osherenko (1995; 2001) has argued more forcefully for the implementation of title to land as a viable instrument for indigenous cultural survival. She states that "without restructuring property rights [the Nenets population] remains powerless to protect its interests against monopolistic and unreformed oil and gas enterprises" (Osherenko, 1995). At present the land is officially owned by the state but is mainly managed by the local authorities, which provide it for use to the sovkhoz (Osherenko, 1995). However in Russia, even if it were granted, title to land is probably not a guarantee against future problems given the limited experience so far with implementation coupled with a weak overall juridical system. The recent trial of entitlements on Yamal during the 1990s was not encouraging. According to Stammler (2005), legal land titles were often granted "for a limited time, which sometimes was not worth the investment in obtaining it." As was demonstrated with ANCSA in Alaska, merely giving people title to valuable land and shares in new enterprises does not ensure that they understand either how to exercise their own rights or to retain them for future generations (Young, 1992).

In the North American petroleum industry, the emphasis is on legal rights and profit sharing. Alaska Natives and Canadian First Nations faced a steep learning curve in the early years of massive development but are now generally able to hold their own in the media, the courts and the marketplace. In contrast, the Yamal Nenets, whose legal rights are being defined piecemeal as part of an ongoing process, realize they are not likely to profit directly in a monetary sense. There are certain inducements and subsidies for people who choose to leave the tundra and take up residence in town. Indeed, relatively minor subsidies to maintain herding in the tundra, such as health care, trade (including barter) and meat distribution are important, even if some are not considered absolutely essential. However, there are no large transfer payments or direct leases to indigenous entities with full or partial title to the land, as has been the norm in North America (Osherenko, 2001). In the Yamal-Nenets and Khanty-Mansisk Autonomous Okrugs, some herders do own shares obtained in the form of vouchers, but there is no equivalent system in the neighboring Nenets Autonomous Okrug (NAO) to the West. To date there are small-scale models under discussion for some form of profit sharing but nothing substantive (F. Stammler, 2006, personal communication).

In their much shorter analysis, Chance and Andreeva (1995) review the "serious problems" related to large-scale petroleum development in northern Alaska and on the Yamal Peninsula. They provide a list of measures they believe would assist in achieving an "equitable" outcome. The series of six recommendations is wide-ranging and complex but can be summarized as follows:

(1) Undertake comparative multidisciplinary and multicultural studies to encourage new modes of thinking about sustainable resource development.

(2) Involve indigenous representatives in policy, planning, implementation and evaluation of every large-scale development in areas where they reside.

(3) Pay more attention to resolving existing environmental crises by promoting social changes now rather than relying on hoped for technological changes in the future.

(4) Internalize so-called "externalities" associated with natural resource development to better reflect the true costs of development.

(5) Bridge the intellectual divide separating physical, biological and social scientists, perhaps via joint work on environmental impact assessments.

(6) Analyze in depth those aspects of the political economy that contribute to the present harm.

Points 1, 5 and 6 are essentially recommendations for future policy-relevant research. The others amount to prescriptive policy bullets. From a Russian perspective, they perhaps reflect a western sense of, and faith in, the basic utility of policy recommendations in the first place. Of these, only portions of Point 2 lie within the realm of possibility in the foreseeable future, at least in the Yamal-Nenets region. Points 3 and 4 rub up directly against the potent mix of hubris and paternalism that continues to characterize not only Russian administration, but also Russian anthropology, as described with contemporary examples by Stammler (2005). The second point comes down to active participation by Nenets in regional planning and assessment, a recommendation that is agreed upon by other academics (Golovnev & Osherenko, 1999; Zen'ko, 2004; Stammler, 2005). By insisting on indigenous involvement in all phases of "every large-scale development where they reside," the recommendation probably stands little chance of full implementation. This is unlikely to change even if western partners are eventually allowed to enter into joint ventures on Yamal, which they currently are not.

In a third analysis, Zen'ko (2004) also identifies problems confronting nomadic Nenets on the Yamal Peninsula. She fears that "the traditional economy is losing even the possibility of existence on a par with its industrial neighbors." According to her, feasibility studies and environmental impact statements do not fully take into account all likely sources of environmental damage. Forbes (1995) pointed out that in addition to the direct impacts of roads and road building on Yamal, there are important indirect or cumulative impacts such as blowing sand and dust from the roads themselves, as well the quarries used to provide construction. Some of these problems were plainly foreseen by the Nenets themselves, such as the introduction of feral dogs to the region as they escape from or are abandoned by oil field workers. Nonetheless, these concerns were not acted upon (Zen'ko, 2004). Another ongoing issue is the apparent unwillingness to actively reclaim or even to clean up used and abandoned sites. High concentrations of garbage and petrochemicals in and around old drill rigs, quarries and along transport corridors present persistent dangers for herders whose migration routes intersect with areas of prior or ongoing activity (Fig. 2).

Zen'ko (2004) recommends creation of an "integrated development program" specifically for the Yamal Peninsula. Such a program would account for its "geographic, historical, demographic and other features" and would "enable development of local law making, to create local legal statutes for the Yamal-Nenets Autonomous District." To western ears, such a prescription may sound so vague as to be meaningless. Yet as she reports, and my colleagues and I have similarly observed during recent fieldwork on Yamal (2004–2007), the actual "demands" of Nenets are so moderate, focused and reasonable that they are agreed to in principle by some of the directors responsible for developing the Yamal Peninsula. These include: (1) complete and timely reclamation of the lands used during the technical work that are not industrial and have no facilities on them; (2) establishing and protecting corridors for movement between camps by people and reindeer herds (Zen'ko, 2004).

The first of these can and should be implemented automatically and unilaterally by the companies responsible for the lease at the time of the development. Unfortunately, as companies have come and gone over the years, overlapping layers of damage, ownership and clean-up responsibilities have become a complicated mess. This was one of the reasons behind the withdrawal of Amoco from Yamal in 1996. The tangled legal web is especially problematic in sectors where development began early and proceeded either continuously or in phases over the last few decades. As some western companies participating in joint ventures have found, even when they are willing to clean up areas damaged by earlier enterprises, there may be legal obstacles delaying or preventing them from doing so.

Fig. 2. Nenets Women Corralling Reindeer in the Vicinity of an Abandoned Drill Rig During Summer Migrations on Yamal Peninsula, July 2005. Herders try to avoid letting animals get too close to such sites because of the rusty metal, broken glass and petro-chemicals that can remain on the ground. when an animal injures a hoof during migration it runs a high risk of becoming infected. If this happens, adult animals tend to be slaughtered since they can no longer keep up with the herd. Herders usually try to treat young calves with leg or hoof wounds and it is not uncommon to see them being transported and fed on special sledges until they can walk properly again. Photo: B. C. Forbes.

The second demand is more difficult because it requires active and sustained bi- or multilateral engagement with herders and indigenous political representatives. Real or meaningful consultation is not easy in practice. There are many obstacles – legal, logistical and attitudinal. Certainly the urgency attached to developing the gas and oil deposits is an important issue, since it drives all other concerns, for better or worse. Yet it is completely unrealistic to expect rapid progress on the issues presented here. As a goal, equity must be seen as a long-term process. Zen'ko (2004) does not call for participatory management by name, but the protection of corridors for movement between camps by people and reindeer herds certainly requires it.

Finally, in a monograph-length analysis of the Yamal Nenets, Stammler (2005) covers virtually all aspects of nomadic reindeer herders' engagement with their own world as well as the "outside" world. His treatment encompasses, but is far from restricted to, their multifaceted responses to the encroachments associated with recent oil and gas activities. He includes a broad historical treatment of their repeated adaptation to major systemic shocks and in doing so effectively documents their resilience, as do Golovnev and Osherenko (1999) in a more circumscribed manner. In his final analysis, Stammler (2005) believes that the nomadic Yamal Nenets will continue to respond flexibly to the changes newcomers continue to usher into their homeland. He adds that the peoples' "competitive spirit will only be a useful force if those who currently hold power agree to continue a fruitful dialogue with the nomads, and genuinely respect the region's current and future identity as one intimately connected with reindeer herding." He also offers the caveat that their "adaptation strategy will depend on whether all actors participating in the development will build on the positive experience of the last century and respect each others' conditions for maintaining flexibility."

SUMMARY AND CONCLUSION

Returning to the broader global context, climate change represents a classic multiscale problem in that it is characterized by infinitely diverse actors, multiple stressors and multiple timescales (Adger, 2006). It has been suggested that climate change impacts will burden most those populations that are already vulnerable to climate extremes, and so bear the brunt of projected (and increasingly observed) changes that are attributable to global climate change (Berkes & Jolly, 2001; Krupnik & Jolly, 2002; ACIA, 2005; Adger, 2006). Yet a key challenge pointed out by prior research on social–ecological systems is the need to match the scale of problems and the social and governance mechanisms devised to cope with them (Folke, 2006; Young et al., 2006). Similar challenges have been exposed in ecology and conservation biology, where the importance of scale and the human dimension has been regularly underestimated in the past (Noss, 1992; Mascia et al., 2003).

Nomadic pastoralists who have endured for centuries, the Yamal Nenets in some ways exemplify resilience. They have successfully adapted to the sequential forms of paternalism from Tsarist through post-Soviet times. As first-hand witnesses to capricious decision-making, they recognize better than all but the most dedicated and open-minded scientists that bridges from the local level to actual policy makers are inadequate or entirely absent. Petroleum development is currently going ahead at full speed, with or without meaningful consultation. To secure a future for reindeer nomadism, mutual coexistence with oil and gas activities is not only possible but also essential, and therefore head-on engagement is the only way forward. In a western context, with recourse to legal action, the situation would be considered ripe for submission to the courts for formal attempts at conflict resolution. Just as there is no current legal recourse, there is no place at all for militant activist or environmentalist tactics, which would conflict with Nenets culture and their historic ability to cooperate with authorities during Soviet times. There is little choice but to work within the current system, albeit one that is characterized by a high degree of dynamism on the one hand and extreme conservatism on the other.

According to several authors in the preceding section, equity and "flexibility" or resilience seem to lie in shared decision making with regard to, among other things, secure access for migrations and maintaining a viable environment. Gray areas encompass matters such as poaching. Illicit hunting and fishing are predictable byproducts of improved mobility to and within remote areas that function as home to economically and/or nutritionally valuable species (Thomassen et al., 1999), but illegal or excessive harvesting by anyone – either indigenous or non-indigenous – must be controlled if the ecosystems are to remain functional. It is easy to forget that even without poaching, certain species of fish and other wildlife can and do undergo significant population fluctuations. This basic tenet of arctic biology has been recognized for decades by scientists (Vibe, 1967), and certainly for much longer by arctic indigenous peoples. Even if the underlying mechanisms are not yet fully understood, in some cases because the cycles occur over long timescales in remote areas well beyond the reach of the "scientific" record, these patterns clearly exist within the indigenous oral record (Ferguson, Williamson, & Messier, 1998; Schneider, Kielland, & Finstad, 2005). For this reason, indigenous observations should be given careful consideration as we move into a new and uncertain phase of rapid change in the Arctic (Huntington et al., 2004). The whole concept of resilience is based on a tacit acceptance of periodic systemic shocks, even if these may not be predictable.

Most authors suggest that Nenets' adaptive capacity is high, but all agree that there must be some accommodation on the part of the government and oil and gas enterprises, a precondition that is described at length by Stammler and Wilson (2006) as "the enabling environment" for dialogue between indigenous peoples, companies and the states. As long as policies are designed and implemented in a strictly top-down manner, we can expect problems to continue. As it is, policy is rarely flexible enough to encompass the level of heterogeneity found in northern social–ecological systems (Bolshakov & Klokov, 2000; Zen'ko, 2004). In the Yamal, as well as in the NAO, some brigades face immediate threats to their future, such as major withdrawals of critical pasturelands, while their neighbors function relatively normally and foresee few serious challenges to their survival. The Nenets themselves are guardedly optimistic (Stammler & Forbes, 2006) and this is an important prerequisite for securing their future on the tundra, but not enough to sustain them.

The heterogeneity of Nenets social–ecological systems means that even neighboring brigades can experience radically different projected and actual trajectories of development and associated impacts. Blanket rules are useful for mitigation strategies, such as enforcing reclamation, poaching and restricting off-road vehicle use, but for consultation, one size does not fit all (Stammler & Wilson, 2006). The balance of power is far from equal, and the more powerful players see little advantage in engaging with the less powerful ones. Reindeer herders do not expect to dictate where and on what schedule development goes ahead, but really do need to be accommodated if pipelines, quarries, railways and roads continue to encroach upon and increasingly fragment their ancestral territories. The rebellions of the 1930s illustrate that flexible adaptation is not the only strategy Nenets have deployed in the past (Golovnev & Osherenko, 1999). Since their leaders, both men and women, are holding positions in the regional Duma and in Russia's Association of Indigenous Peoples of the North (RAIPON), they may well find ways to deal with Gazprom.

With a situation like that in modern Russia, an analysis such as this is the easy part. Coming after 16 years of working there, enabling responsible social and environmental policy is far more challenging. We must accept without question the errors made in Soviet times and learn from them. Only then is it possible to encourage adaptive capacity for the future. Demography is key (Pika & Bogoyavlensky, 1995). Young people must want to live and remain in a place … period. Even in highly developed countries like Norway, huge subsidies are currently unable to sustain marginal communities, such as fishing villages along the northern coast of the Barents Sea, where young people continue to leave for education and employment opportunities in places like Tromsø, Oslo and beyond. While state subsidies can and do assist rural populations with a strong desire to live as such, a sense of place cannot be legislated, just as a way of knowing about the land cannot be learned in a classroom. Unless young people see a future in it, the Yamal tundra, not unlike any other place, is doomed to be gradually depopulated. The importance of youth retention is something that transcends boundaries in the Arctic. As Zen'ko (2004) puts it, "education emphasizing tundra survival skills is seen as a hedge or life preserver in a whirlpool of socioeconomic injustice." While I do not have any official statistics to support them, my own observations on Yamal in the post-Soviet era indicate that there is a rough balance of gender and age in the migratory population and that young people continue to choose to return to the tundra upon completion of their school and military service obligations. This in and of itself is highly likely to be a strong indicator of resilience. At the same time, the ongoing presence of a sizable nomadic Nenets population may serve as an incentive for the state to maintain a viable social and ecological system. This is because those in power in Yamal have understood that intact nomadic reindeer herding as a regional identity marker increases the status of the Okrug (Stammler, 2005).

The lack of meaningful consultation is hardly unique to Russia. We saw that for reindeer management in northernmost Fennoscandia, power relations are similarly skewed in relation to forestry, hydropower, mining and tourism (Forbes et al., 2004a, 2006). Even among older and ostensibly enlightened western democracies of northernmost Europe, the analysis of modern reindeer management has exposed significant disparities in power relations between the respective states and indigenous interests (Müller-Wille et al., 2006; Hukkinen et al., 2006). In North America, several observers have identified a lack of trust in co-management boards, even when community "users" serve as board members or are otherwise party to scientific "predictions" concerning wildlife monitoring and management (Kruse et al., 1998; Nadasdy, 2003, in press; Moller et al., 2004). Yet the literature of resilience as it pertains to high-latitude social–ecological systems is dominated by cases of marine mammal and caribou co-management from North America (Berkes, 1998; Berkes & Jolly, 2001; Kendrick, 2003a).

The point is not to disparage the North American-based literature and examples of northern resource management regimes, but to point out their limitations in terms of applicability to the rest of the Arctic. Co-management, for one, has been hailed widely as a success in North America (Nadasdy, 2003, in press). Certainly there have been some positive outcomes in Canada and Alaska, but there are still problems in transferring these models overseas. Robards and Alessa (2004) rightly point out that "the significant challenge of maintaining equity and resilience of remote communities, within and outside the Arctic, will necessitate incorporating localized cultural values and decision-making processes that fostered prior community existence with (data from) western interdisciplinary research." However, in Europe and Russia, the institutional barriers to participatory modes of decision-making in resource management mean that North American standards of equity are not viable models for the time being. A North American-style legal system of title to land is also neither likely to be implemented, nor a caveat to solve the panoply of problems surrounding contemporary nomadic reindeer herding as practiced by the Yamal Nenets. On the other hand, secure access to traditional lands for herding, hunting, fishing is an essential component of maintaining a viable human population within the region. Self-determination is also likely to play an important role in leveraging some form of property rights, but to what extent will depend in part on what course the current and future Nenets leaders choose to navigate.

Given the closely guarded manner in which the Russian state holds the reins of development, via its functional control of Gazprom, multinationals are not likely to be invited to participate in the development of Yamal in the foreseeable future. As a result, there is actually little that sympathetic westerners can do to actively support the Nenets' struggle for mutual coexistence. If such joint ventures eventually do come about, they can and should be carefully monitored to ensure responsible behavior. What are important in the meantime is that the Russian state itself works to make mutual coexistence feasible and that the migratory tundra Nenets provide a potent example of long-term resilience in a social–ecological system that bears little resemblance to those featured in the western literature to date.

NOTES

1. Yamal-Nenets Autonomous Okrug (YNAO) belongs to the West Siberian economic region and Ural Federal district. YNAO is an independent unit of the Tyumen Oblast and lies in the extreme north of the West Siberian lowland.

2. Fennoscandia is a geographic term based on linguistics that encompasses the Scandinavian-speaking countries (Norway, Sweden, Denmark, Faroe Islands, Iceland) and Finland. The Finnish language derives from the Finno-Ugric group.

ACKNOWLEDGMENTS

My research on the Yamal Peninsula in the 1990s was funded from several different sources. These included the National Science Foundation (U.S.), the National Geographic Society, NATO's Scientific and Environmental

Affairs Division and the Russian Academy of Sciences. Exceptionally useful support from the Academy of Finland has been provided to the ENSINOR project under the auspices of their "Russia in Flux" program during the years 2004–2007. Additional support came from the National Science Foundation Office of Polar Programs and the National Aeronautics and Space Administration through the Northern Eurasian Earth Science Partnership Initiative. I am especially grateful to my dear colleagues Nina Meschtyb, Florian Stammler, Timo Kumpula and Anu Pajunen for their warm companionship, consistent professionalism and endlessly stimulating discussions on our many and varied and sojourns to Yamal. Dr. Stammler, Gail Osherenko and an anonymous reviewer generously provided constructive comments on the manuscript.

REFERENCES

ACIA. (2005). *Arctic climate impact assessment.* Cambridge: Cambridge University Press.
AHDR. (2004). *Arctic human development report.* Akureyri: Stefansson Arctic Institute.
Adger, W. N. (2006). Vulnerability. *Global Environmental Change, 16,* 268–281.
Armstrong, T., Rogers, G., & Rowley, G. (1978). *The circumpolar North.* New York: Methuen, Inc.
Bankes, N. (2004). Legal systems. In: *Arctic human development report* (pp. 101–118). Akureyri: Stefansson Arctic Institute.
Baskin, L. M. (2000). Reindeer husbandry/hunting in Russia in the past, present and future. *Polar Research, 19,* 23–29.
Berger, T. R. (1985). *Village journey: The report of the Alaska Native Review Commission.* New York: Hill and Wang.
Berkes, F. (2007). Adaptive co-management and complexity: Exploring the many faces of co-management. In: D. Armitage, F. Berkes, & N. Doubleday (Eds.), *Adaptive co-management* (pp. 19–37). Vancouver: University of British Columbia Press.
Berkes, F. (1998). Indigenous knowledge and resource management systems in the Canadian subarctic. In: F. Berkes, C. Folke & J. Colding (Eds), *Linking social and ecological systems* (pp. 98–127). Cambridge: Cambridge University Press.
Berkes, F. (1999). *Sacred ecology: Traditional ecological knowledge and resource management.* Philadelphia: Taylor & Francis.
Berkes, F., Bankes, N., Marschke, M., Armitage, D., & Clark, D. (2005). Cross-scale institutions and building resilience in the Canadian North. In: F. Berkes, R. Huebert, H. Fast, M. Manseau & A. Diduck (Eds), *Breaking ice: Renewable resource and ocean management in the Canadian North* (pp. 225–247). Calgary: University of Calgary Press.
Berkes, F., & Folke, C. (2002). Back to the future: Ecosystem dynamics and local knowledge. In: L. Gunderson & C. S. Holling (Eds), *Panarchy: Understanding transformations in human and natural systems* (pp. 121–146). Washington, DC: Island Press.

Berkes, F., & Jolly, D. (2001). Adapting to climate change: Social–ecological resilience in a Canadian western Arctic community. *Conservation Ecology, 5*(2), 18.

Bolshakov, N. N., & Klokov, K. B. (2000). Protected areas in the North of Russia and problems of northern minorities. In: B. S. Ebbinge, Yu. L. Mazourov & P. S. Tomkovich (Eds), *Heritage of the Russian Arctic* (pp. 572–577). Moscow: Ecopros Publishers.

Caulfield, R. A. (2000). Political economy of renewable resources in the Arctic. In: M. Nuttall & T. V. Callaghan (Eds), *The Arctic: Environment, people, policy* (pp. 485–513). Amsterdam: Harwood Academic Publishers.

Chance, N. A., & Andreeva, E. N. (1995). Sustainability, equity, and natural resource development in northwest Siberia and arctic Alaska. *Human Ecology, 23*, 217–240.

Chapin, F. S., III., Angelstam, P., Apps, M., Berkes, F., Folke, C., Forbes, B. C., Juday, G., & Peterson, O. (2004). Vulnerability and resilience of high-latitude ecosystems to environmental and social change. *Ambio, 33*, 344–349.

Chapin, F. S., III., Hoel, M., Carpenter, S. R., Lubchenco, J., Walker, B., Callaghan, T. V., Folke, C., Levin, S. A., Mäler, K.-G., Nilsson, C., et al. (2006a). Building resilience and adaptation to manage arctic change. *Ambio, 35*, 1–5.

Chapin, F. S., III., Robards, M. D., Huntington, H. P., Johnstone, J. F., Trainor, S. F., Kofinas, G. P., Ruess, R. W., Fresco, N., Natcher, D. C., & Naylor, R. L. (2006b). Directional changes in ecological communities and social–ecological systems: A framework for prediction based on Alaskan examples. *American Naturalist, 168*, 36–49.

Chapin, F. S., III., Rupp, T. S., Starfield, A., M DeWilde, L., Zavaleta, E. S., Fresco, N., Henkelman, J., & McGuire, A. D. (2003). Planning for resilience: Modeling change in human-fire interactions in the Alaskan boreal forest. *Frontiers in Ecology and Environment, 1*(5), 255–261.

Damas, D. (Ed.) (1984). *Handbook of North American Indians* (Vol. 5. Arctic). Washington, DC: Smithsonian Institution.

Fedorova, N. (Ed.) (1998). *Gone to the hills: Culture of the coastal residents of the Yamal Peninsula during the Iron Age (in Russian)*. Ekaterinburg: History and Archaeology Institute.

Felix, N. A., Raynolds, M. K., Jorgenson, J. C., & DuBois, K. E. (1992). Resistance and resilience of tundra plant communities to disturbance by winter seismic vehicles. *Arctic and Alpine Research, 24*, 69–77.

Ferguson, M. A. D., & Messier, F. (1997). Collection and analysis of traditional ecological knowledge about a population of arctic tundra caribou. *Arctic, 50*, 17–28.

Ferguson, M. A. D., Williamson, R. G., & Messier, F. (1998). Inuit knowledge of long-term changes in a population of arctic tundra caribou. *Arctic, 51*, 201–219.

Flanders, N. E. (1989). The ANCSA amendments of 1987 and land management in Alaska. *Polar Record, 25*, 315–322.

Folke, C. (2006). Resilience: The emergence of a perspective for social–ecological systems analyses. *Global Environmental Change, 16*, 253–267.

Folke, C., Berkes, F., & Colding, J. (2003). Synthesis: Building resilience and adaptive capacity in social–ecological systems. In: F. Berkes, J. Colding & C. Folke (Eds), *Navigating social–ecological systems* (pp. 352–387). Cambridge: Cambridge University Press.

Folke, C., Carpenter, S., Elmqvist, T., Gunderson, L., Holling, C. S., Walker, B., Bengtsson, J., Berkes, F., Colding, J., & Danell, K. etal. (2002). *Resilience and sustainable development:*

Building adaptive capacity in a world of transformations. Stockholm: Environmental Advisory Council.

Forbes, B. C. (1995). Tundra disturbance studies, III: Short-term effects of Aeolian sand and dust, Yamal Region, northwest Siberia, Russia. *Environmental Conservation, 22,* 335–344.

Forbes, B. C. (2005a). Wilderness. In: M. Nuttall (Ed.), *Encyclopedia of the Arctic* (pp. 2185–2187). New York: Routledge.

Forbes, B. C. (2005b). Conservation. In: M. Nuttall (Ed.), *Encyclopedia of the Arctic* (pp. 427–432). New York: Routledge.

Forbes, B. C. (2005c). Reindeer. In: M. Nuttall (Ed.), *Encyclopedia of the Arctic* (pp. 1750–1752). New York: Routledge.

Forbes, B. C. (2006). The challenges of modernity for reindeer management in northernmost Europe. In: B. C. Forbes, et al. (Eds), *Reindeer management in northernmost Europe: Linking practical and scientific knowledge in social–ecological systems* (pp. 11–25). Berlin: SpringerEcological Studies, 184.

Forbes, B. C., Bölter, M., Müller-Wille, L., Hukkinen, J., Müller, F., Gunslay, N., & Konstantinov, Y. (Eds.), (2006). *Reindeer management in northernmost Europe: Linking practical and scientific knowledge in social–ecological systems.* Berlin: Springer, *Ecological Studies, 184.*

Forbes, B. C., Fresco, N., Shvidenko, A., Danell, K., & Chapin, F. S., III. (2004b). Geographic variations in anthropogenic drivers that influence the vulnerability and resilience of high latitude social–ecological systems. *Ambio, 33,* 377–382.

Forbes, B. C., & Kofinas, G. (Eds.), (2000). The human role in reindeer and caribou grazing systems. *Polar Research, 19*(1), 1–142.

Forbes, B. C., Monz, C., & Tolvanen, A. (2004a). Tourism ecological impacts in terrestrial polar ecosystems. In: R. Buckley (Ed.), *Environmental impacts of ecotourism* (pp. 155–170). Oxfordshire: CAB International.

Golovnev, A. V., & Osherenko, G. (1999). *Siberian survival: The Nenets and their story.* Ithaca, NY: Cornell University Press.

Harrington, F. (2005). Caribou. In: M. Nuttall (Ed.), *Encyclopedia of the Arctic* (pp. 318–319). New York: Routledge.

Helle, T., & Kojola, I. (2006). Population trends of semi-domesticated reindeer in Fennoscandia – evaluation of explanations. In: B. C. Forbes, M. Bölter, L. Müller-Wille, J. Hukkinen, F. Müller, N. Gunslay & Y. Konstantinov (Eds), *Reindeer management in northernmost Europe: Linking practical and scientific knowledge in social–ecological systems* (pp. 319–339). Berlin: SpringerEcological Studies, 184.

Helm, J. (Ed.) (1981). *Handbook of North American Indians* (Vol. 6. Subarctic). Washington, DC: Smithsonian Institution.

Holling, C. S., & Gunderson, L. (2002). Resilience and adaptive cycles. In: L. Gunderson & C. S. Holling (Eds), *Panarchy: Understanding transformations in human and natural systems* (pp. 25–62). Washington, DC: Island Press.

Hoyos, C., Bream, R., Harvey, F., & McNulty, S. (2006). BP could learn from Exxon example. *Financial Times, August 12–13,* 9.

Hukkinen, J., Müller-Wille, L., Aikio, P., Heikkinen, H., Jääskö, O., Laakso, A., Magga, H., Nevalainen, S., Pokuri, O., Raitio, K., et al. (2006). Development of participatory institutions for reindeer management in Finland: A diagnosis of deliberation, knowledge integration and sustainability. In: B. C. Forbes, et al. (Eds), *Reindeer management in northernmost Europe: Linking practical and scientific knowledge in social–ecological systems* (pp. 47–71). Berlin: Springer, *Ecological Studies, 184.*

Huntington, H., Callaghan, T., Fox, S., & Krupnik, I. (2004). Matching traditional and scientific observations to detect environmental change: A discussion on arctic terrestrial ecosystems. *Ambio Special Report, 13,* 18–23.

Huntington, H. P. (2000). Using traditional ecological knowledge in science: Methods and applications. *Ecological Applications, 10,* 1270–1274.

Ingold, T. (1980). *Hunters, pastoralists and ranchers.* Cambridge: Cambridge University Press.

Janssen, M. A., Schoon, M. L., Ke, W., & Börner, K. (2006). Scholarly networks on resilience, vulnerability and adaptation within the human dimensions of global environmental change. *Global Environmental Change, 16,* 240–252.

Jentoft, S. (Ed.) (1998) *Commons in a cold climate: Coastal fisheries and reindeer pastoralism in North Norway.* Paris: Parthenon Publishers.

Jernsletten, J.-L., & Beach, H. (2006). The challenges and dilemmas of concession reindeer management in Sweden. In: B. C. Forbes, et al. (Eds), *Reindeer management in northernmost Europe: Linking practical and scientific knowledge in social–ecological systems* (pp. 95–116). Berlin: Springer, *Ecological Studies, 184.*

Karlstad, S. (1998). Institutional theory, co-management and sustainable development in Saami reindeer pasture commons – critical factors for a robust system of local management. In: S. Jentoft (Ed.), *Commons in cold climate: Coastal fisheries and reindeer pastoralism in North Norway* (pp. 17–39). Paris: Parthenon Publishers.

Karppi, K., & Eriksson, J. (Eds). (2002). *Conflict and cooperation in the North.* Umeå: Norrlands University Press.

Kendrick, A. (2002). Caribou co-management: Realizing conceptual differences. *Rangifer, 13,* 7–13.

Kendrick, A. (2003a). Caribou co-management in northern Canada: Fostering multiple ways of knowing. In: F. Berkes, J. Colding & C. Folke (Eds), *Navigating social–ecological systems* (pp. 241–267). Cambridge: Cambridge University Press.

Kendrick, A. (2003b). The flux of trust: Caribou co-management in northern Canada. *Environments, 31,* 43–59.

Kendrick, A. (2003c). *Caribou co-management and cross-cultural knowledge sharing.* Ph.D. Thesis, University of Manitoba, Winnipeg.

Khorolya, D. (2002). Reindeer husbandry in Russia. In: S. Kankaanpää (Ed.), *The 2nd World Reindeer Herders' Congress Anár 2001* (pp. 40–42). Rovaniemi: University of Lapland, *Arctic Centre Reports, 36.*

Kitti, H., Gunslay, N., & Forbes, B. C. (2006). Defining the quality of reindeer pastures: The perspectives of Sámi reindeer herders. In: B. C. Forbes, et al. (Eds), *Reindeer management in northern most Europe: Linking practical and scientific knowledge in social–ecological systems* (pp. 141–165). Berlin: Springer, *Ecological Studies, 184.*

Klein, D. R., Moorehead, L., Kruse, J., & Braund, S. R. (1999). Contrasts in use and perceptions of biological data for caribou management. *Wildlife Society Bulletin, 27,* 488–498.

Klokov, K. (2000). Nenets reindeer herders on the lower Yenisei River: Traditional economy under current conditions and responses to economic change. *Polar Research, 19,* 39–47.

Krupnik, I. (1993). *Arctic adaptations: Native whalers and reindeer herders of northern Eurasia.* Hanover: University Press of New England.

Krupnik, I., & Jolly, D. (Eds). (2002). *The earth is faster now: Indigenous observations of arctic environmental change.* Fairbanks: ARCUS.

234   BRUCE C. FORBES

Kruse, J., Klein, D., Braund, S., Moorehead, L., & Simeon, B. (1998). Co-management of natural resources: A comparison of two caribou management systems. *Human Organization, 57*, 447–458.

Martens, H. (1995). *Revegetation research western Siberia: Year 4.* Report prepared for Amoco Production Co., Harvey Martens & Assoc., Inc., Calgary.

Martens, H., Magomedova, M., & Morozova, L. (1996). *Rangeland studies in the Bovanenkovo proposed development area: Year 3.* Report prepared for Amoco Eurasia Production Co. Harvey Martens & Assoc., Inc., Calgary and Institute of Flora and Fauna, Ekaterinburg.

Mascia, M., Brosius, J. P., Dobson, T., Forbes, B. C., Nabhan, G., & Tomforde, M. (2003). Conservation and the social sciences. *Conservation Biology, 17*, 649–650.

Moller, H., Berkes, F., Lyver, P. O., & Kislalioglu, M. (2004). Combining science and traditional ecological knowledge: Monitoring populations for co-management. *Ecology and Society, 9*(3), 2.

Mufson, S. (2006). Small leaks in Alaska – big trouble for BP. *Guardian Weekly, August 18–24*, 27.

Mysterud, A. (2006). The concept of overgrazing and its role in management of large herbivores. *Wildlife Biology, 12*, 129–141.

Müller-Wille, L., Heinrich, D., Lehtola, V.-P., Aikio, P., Konstantinov, Y., & Vladimirova, V. (2006). Dynamics in human-reindeer relations: Reflections on prehistoric, historic and contemporary practices in northernmost Europe. In: B. C. Forbes, et al. (Eds), *Reindeer management in northernmost Europe: Linking practical and scientific knowledge in social–ecological systems* (pp. 27–45). Berlin: Springer, *Ecological Studies, 184.*

NRC. (2003). *Cumulative environmental effects of oil and gas activities on Alaska's north slope.* Washington, DC: National Research Council.

Nadasdy, P. (2003). Re-evaluating the co-management success story. *Arctic, 56*, 367–380.

Nadasdy, P. (2007). Adaptive co-management and the gospel of resilience. In: D. Armitage, F. Berkes, & N. Doubleday (Eds), *Adaptive co-management: Collaboration, learning and multi-level governance* (pp. 208–227). Vancouver: University of British Columbia Press.

Nelson, R. K. (1969). *Hunters of the northern ice.* Chicago: University of Chicago Press.

Noss, R. (1992). Issues of scale in conservation biology. In: P. L. Fiedler & S. K. Jain (Eds), *Conservation biology: The theory and practice of nature conservation and management* (pp. 239–250). New York: Chapman & Hall.

Nuttall, M., Berkes, F., Forbes, B. C., Kofinas, G., Vlassova, T., & Wenzel, G. (2005). Hunting, herding, fishing and gathering. In: *Arctic climate impact assessment* (pp. 649–690). Cambridge: Cambridge University Press.

Olson, D. M., & Dinerstein, E. (1998). The Global 200: A representation approach to conserving the earth's most biologically valuable ecoregions. *Conservation Biology, 12*, 502–515.

Olsson, P., Folke, C., & Berkes, F. (2004). Adaptive co-management for building resilience in social–ecological systems. *Environmental Management, 34*, 75–90.

Osherenko, G. (1995). Property rights and transformation in Russia: Institutional change in the Far North. *Europe–Asia Studies, 47*, 1077–1108.

Osherenko, G. (2001). Indigenous rights in Russia: Is title to land essential for cultural survival? *The Georgetown International Environmental Law Review, 13*, 695–733.

O'Carroll, E., & Elliot, M. (2005). *Greenland and the Arctic* (2nd ed.). London: Lonely Planet Publications.

Peters, E. J. (2003). Views of traditional ecological knowledge in co-management bodies in Nunavik, Quebec. *Polar Record, 39,* 49–60.

Pika, A., & Bogoyavlensky, D. (1995). Yamal Peninsula: Oil and gas development and problems of demography and health among indigenous populations. *Arctic Anthropology, 32,* 61–74.

Regalado, A. (2005). Global warring: In climate debate, the 'hockey stick' leads to a face off. *The Wall Street Journal, February 14,* 1.

Reynolds, J. F., & Tenhunen, J. D. (1996). Ecosystem response, resistance, resilience, and recovery in arctic landscapes: Introduction. In: J. F. Reynolds & J. D. Tenhunen (Eds), *Landscape function and disturbance in arctic tundra* (pp. 3–18). Berlin: Springer.

Robards, M., & Alessa, L. (2004). Timescapes of community resilience and vulnerability in the circumpolar North. *Arctic, 57,* 415–427.

Sandström, P., Pahlén, T. G., Edenius, L., Tømmervik, H., Hagner, O., Hemberg, L., Olsson, H., Baer, K., Stenlund, T., Brandt, L. J., & Egberth, M. (2003). Conflict resolution by participatory management: remote sensing and GIS as tools for communicating land-use needs for reindeer herding in northern Sverige. *Ambio, 32,* 557–567.

Schneider, W., Kielland, K., & Finstad, G. (2005). Factors in the adaptation of reindeer herders to caribou on the Seward Peninsula, Alaska. *Arctic Anthropology, 42,* 36–49.

Smit, B., & Pilifosova, O. (2001). Adaptation to climate change in the context of sustainable development and equity. In: *Intergovernmental panel on climate change, third assessment report* (pp. 878–912). Cambridge: Cambridge University Press.

Stammler, F. (2002). Success at the edge of the land: Present and past challenges for reindeer herders of the West-Siberian Yamal-Nenets autonomous Okrug. *Nomadic Peoples, 6,* 51–71.

Stammler, F. (2005). *Reindeer nomads meet the market: Culture, property and globalization at the 'end of the land'.* Berlin: Lit Verlag.

Stammler, F., & Forbes, B. C. (2006). Oil and gas development in the Russian Arctic: West Siberia and Timan-Pechora. *IWGIA Newsletter, 2-3/06,* 48–57.

Stammler, F., & Wilson, E. (2006). Dialogue for development: An exploration of relations between oil and gas companies, communities and the state. In: F. Stammler, E. Wilson (Eds.), *Oil and gas industry, local communities and the state. Sibirica, 5*(2), 1–42.

Talley, I. (2006). Arctic harshness hinders search for oil. *The Wall Street Journal, July 11,* A11.

Thomassen, J., Dallmann, W., Isaksen, K., Khlebovich, V., & Wiig, Ø. (1999). *Evaluation of INSROP valued ecosystem components: Protected areas, indigenous people, domestic reindeer and wild reindeer.* INSROP Working Paper No. 162-1999, II.5.10. Fridtjof Nansen Institute, Lysaker, Norway.

Treseder, L. (1999). The evolution and status of wildlife co-management in Canada. In: L. Treseder & J. Honda-McNeil (Eds), *Northern Eden: Community based wildlife management in Canada* (pp. 7–18). Edmonton: University of Alberta.

Usher, P. J. (2000). Traditional ecological knowledge in environmental assessment and management. *Arctic, 53,* 183–193.

Vakhtin, N. B. (1992). *Native peoples of the Russian Far North.* London: Minority Rights Group.

Vibe, C. (1967). Arctic animals in relation to climatic fluctuations. *Meddelelser om Grønland, 170*(5).

van der Wal, R. (2006). Do herbivores cause habitat degradation or vegetation state transition? Evidence from the tundra. *Oikos, 114,* 177–186.

WWF. (2006). WWF International Arctic Programme. http://www.panda.org/about_wwf/where_we_work/arctic/offices/index.cfm (accessed on September 11, 2006).

Wenzel, G. W. (1991). *Animal rights human rights: Ecology, economy and ideology in the Canadian Arctic.* Toronto: University of Toronto Press.

Whiteman, G., Forbes, B. C., Niemelä, J., & Chapin, F. S., III. (2004). Bringing high-latitude feedback and resilience into the corporate boardroom. *Ambio, 33*, 371–376.

Wookey, P. A., & Robinson, C. H. (1997). Responsiveness and resilience of high arctic ecosystems to environmental change. *Opera Botanica, 132*, 215–232.

Young, O. R. (1992). *Arctic politics: Conflict and cooperation in the circumpolar North.* Dartmouth: University Press of New England.

Young, O. R., Berkhoutb, F., Gallopin, G. C., Janssen, M. A., Ostrom, E., & van der Leeuw, S. (2006). The globalization of socio-ecological systems: An agenda for scientific research. *Global Environmental Change, 16*, 304–316.

Young, O. R., & Osherenko, G. (1992). Arctic resource conflicts: Sources and solutions. In: O. R. Young (Ed.), *Arctic politics: Conflict and cooperation in the circumpolar North* (pp. 104–125). Dartmouth: University Press of New England.

Zen'ko, M. A. (2004). Contemporary Yamal: Ethncoecological and ethnosocial problems. *Anthropology & Archaeology of Eurasia, 42*, 7–63.

PART IV:
INEQUALITY AND EQUITY AS PREDICTORS OF ENVIRONMENTAL HARM

INEQUALITY IN THE CREATION OF ENVIRONMENTAL HARM: LOOKING FOR ANSWERS FROM WITHIN

Lisa M. Berry

ABSTRACT

To date, many environmental policy discussions consider inequalities between groups (typically by comparing the average or aggregate resource use of one group to another group), but most ignore disproportionalities **within groups**. Disproportionality, as discussed in a small but growing body of work, refers to resource use that is highly unequal among members of the **same** group, and is characterized by a positively skewed distribution, where a small number of resource users create far more environmental harm than "typical" group members. Focusing on aggregated or average impacts effectively treats all members of a group as interchangeable, missing the few "outliers" that actually tend to be responsible for a large fraction of overall resource use. This chapter offers reasons why we should or should not **expect** disproportional production of environmental impacts (from both mathematical and sociological perspectives), looks at empirical evidence of disproportionality, and offers a framework for detecting disproportionality and assessing just how much difference the outliers make. I find that in cases

Equity and the Environment
Research in Social Problems and Public Policy, Volume 15, 239–265
Copyright © 2008 by Elsevier Ltd.
ISSN: 0196-1152/doi:10.1016/S0196-1152(07)15007-9

where the within-group *distribution of resource use is highly dispropor-tionate (characterized by extreme outliers), targeting reduction efforts at the disproportionate polluters can offer opportunities to decrease environmental degradation substantially, at a relatively low cost.*

BACKGROUND

In *State of the World 2006*, Worldwatch Institute asks if there is enough "ecological capacity" for everyone. The answer to this question, in their account, depends on how many resources are consumed by the average world citizen, which, in turn, depends on the average consumption rate within each country. Resource consumption is measured in global hectares, using a technique called Ecological Footprinting (Wackernagel & Rees, 1996). The Ecological Footprint attempts to quantify the amount of resources needed to support an economy by converting the types and quantities of resources used for food, housing, transportation, consumer goods, and services into a single number (the amount of biologically productive land), called the "footprint" of the economy. Worldwatch Institute addresses the vast inequality of resource consumption (footprints) between more and less developed countries as follows:

> The unequal claims on biocapacity become clear when they are analyzed on a per person basis. The average Indian or Chinese footprint is well under the world average of 2.3 global hectares. In contrast, the average Japanese and European each required roughly 4.5 global hectares to support their lifestyles. And the average American is in a separate league entirely, with a footprint of 9.7 global hectares. (Flavin and Gardner, 2006, p. 16)

Importantly, these disproportionate footprints are expressed in terms of demand "per person," or "by each person," but in fact, they actually represent the aggregate, or *total* environmental impact in each country, divided by the number of people in that country (known as the "average" or "arithmetic mean"). The Ecological Footprint is often used for this type of coarse, national-level comparison, because "consumption, production, and trade data are generally compiled at the national level by domestic statistical offices," while "data specific to lower-level political entities such as states, provinces, or cities are generally much harder to come by" (Marcotullio & McGranahan, 2005, p. 817). Still, when measures of resource consumption are aggregated in this way, one of the key stories of consumption inequality fails to be revealed – the story of *within-group* variation. As the following pages will spell out, recent research suggests that this kind of focus on "average" or "total" amounts of resource use paints an inaccurate portrait of how much resource consumption is "necessary" to support a certain standard of living, often *overestimating* the amount of consumption that is associated with a "typical" person, sector, or industry within an economy.

This analysis is presented in four main sections. First, I provide a review of how the resource efficiency of an economy has been characterized in mainstream literature. Next, I draw on sociological and statistical theories to explore whether or not we should *expect* disproportional consumption of resources. Third, I trace the roots of Disproportionality Theory, first proposed by Freudenburg (1997) and Freudenburg and Nowak (2000), and examine empirical evidence, noting that where the *within-group* distribution of resource use is highly disproportionate (characterized by extreme outliers), targeting reduction efforts at the few disproportionate polluters offers an opportunity to decrease resource consumption substantially, at a relatively low cost. Finally, I consider how much of a difference "outliers" make when within-group disproportionality exists, and provide an approach for detecting instances of large inequalities within groups more systematically.

RESOURCE CONSUMPTION: GETTING THE NUMBERS RIGHT

Footprint analysis provides a useful starting point for characterizing resource consumption, but the metric can hide important variations within groups. Just as narrowly focusing on the "world average of 2.3 global hectares" fails to reveal the vast inequalities in consumption rates *between* nations, narrowly focusing on a "United States average of 9.7 global hectares" omits crucial information about the vast inequalities *within* the United States. Preliminary research indicates that within-group inequality of resource consumption (in this case, inequalities within specific countries) may be even greater than between-group inequality (in this case, the inequality between countries). A growing body of research suggests that within-group inequality in resource consumption (whether it be "within-country," "within-industry," or "within a specific sector of an industry where firms are producing comparable and relatively homogeneous products") is characterized by a few actors that are responsible for disproportionately large shares of the resources being used or of the pollution being generated (Freudenburg, 2005, 2006; Nowak & Cabot, 2004;

Nowak, Bowen, & Cabot, 2006). The resultant implication is that focusing on the mathematical "average" tends to *overestimate* the amount of resource use that "typifies" the majority of the people or firms in each group.

This difference between "average" and "typical" may be obscure. According to Webster's New World Dictionary, "average" has two distinct meanings. One is "the numerical result obtained by dividing the sum of two or more quantities by the number of quantities; an arithmetical mean," while the other is more broadly "the usual or normal kind, amount, quality, rate, etc.," or "a number or value that typifies a set of values of which it is a function, as a median or mode (Guralnik, 1986)."

In some cases, the first or mathematical definition of "average" (arithmetic mean) can be roughly equivalent to the notion of "average" being a typical, or representative measure of group performance, but that will not always be the case. If all members of a group consume exactly the same quantity of resources (for example, if each person in the United States were to consume 9.7 global hectares), the "average resource use" would be equal to the "per capita resource use" for every single member of the group. Thus, all members of the group are "interchangeable." In this case, the impact of each member of the group truly would be proportional to the group's total resource consumption divided by the number of members (see Case A, "within-group Proportionality" in Table 1), and using the "arithmetic mean" would accurately reflect the "normal amount" of pollution for the group. However, if within-group consumption is not equal, that is, group members are not "interchangeable" in their consumption rates, the mean may not typify the set of values (see Case B,

Table 1. Hypothetical Distributions of Resource Use among Five U.S. Citizens.

United States Citizen	Ecological Footprint (Global Hectares)	
	Case A: Within-group Proportionality	Case B: Within-group Disproportionality
Citizen #1	9.7	1
Citizen #2	9.7	2
Citizen #3	9.7	2.5
Citizen #4	9.7	3
Citizen #5	9.7	40
Average footprint	9.7	9.7

"within-group Disproportionality" in Table 1). Consider the following hypothetical consumption rates for five United States citizens:

Even though the "average" footprint is the same (9.7 global hectares) for both distributions of resource use in Table 1, the average is far from the "normal" amount of consumption for the Disproportionate case. In this hypothetical example, all citizens but one actually consume far *below* the mean for the group. This example illustrates the statistical concept of a "positive skewed distribution" (an asymmetric probability distribution characterized by a long tail on the right side), as just one person (Citizen #5 in Case B) is responsible for a disproportionate impact on the average consumption for the group.

In statistics, the problem of an "outlier" (a value far away from the center of the distribution) influencing the average of a distribution can be resolved by avoiding the use of the mean, and instead using the median (which is the "middle" number in a group when values are arranged in ascending order, and a more robust measure of central tendency for skewed distributions). For the example set of values, the median is 2.5 global hectares, which is much more "typical" for the five citizens than the value of 9.7. However, simply replacing the mean with the median still fails to draw attention to the most important part of the disproportionate distribution, the single citizen who consumes 40 of the 48.5 global hectares (which is the total consumption for all five citizens in this hypothetical example). Importantly, without actually examining the distribution of resource consumption within each group, the citizens in the Proportionate distribution appear to consume the exact same amount of resources as the citizens in Disproportionate distribution. Clearly, reporting "per-capita" consumption can greatly overestimate the "typical" resource use for disproportionate distributions. In the next section, accordingly, I explore commonly accepted methods for characterizing societies' resource consumption, which are likely to misrepresent the "normal" amount of resource consumption when the underlying distribution is disproportionate.

RESOURCE CONSUMPTION: TENDENCY TO FOCUS ON BETWEEN-GROUP INEQUALITY

The use of the "average," "per-capita," and/or "total" amount, in describing resource consumption, is so pervasive that this author was hard-pressed to find analyses that did *not* employ one or more of these measures. Recent efforts to characterize the exploitation of planet Earth's resources and ecosystem services, such as the *AAAS Atlas of Population and Environment* (Harrison & Pearce, 2001), *Living Planet Report, 2006* (World Wildlife Fund, 2006), and *State of the World, 2006* (Flavin & Gardner, 2006) all report both per-capita and total resource consumption between nations; the latter two publications explicitly use the technique of Ecological Footprinting.

The Ecological Footprint (Wackernagel & Young, 1998) embodies the same basic principles as the "IPAT" model developed 27 years earlier by Paul Ehrlich and John Holdren (Ehrlich & Holdren, 1971), which was later subjected to several revisions and refinements, such as the "STIRPAT" model by York, Rosa, and Dietz (both of which will be discussed below). Both models use three parameters to characterize the resource consumption of a group: the size of the population, the per capita consumption, and the "intensity" of the consumption (which is often considered to be a function of available technology – see Fig. 1 below)

FOOTPRINT AND BIOCAPACITY FACTORS THAT DETERMINE OVERSHOOT

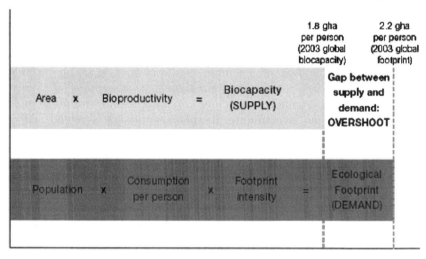

Fig. 1. "Footprint" of Resource Consumption (demand) is Based on Three Parameters: Population, Per-capita Consumption, and Footprint Intensity. Notably Absent is any Measure of the Variance in Per-capita Consumption (World Wildlife Fund, 2006).

The Ecological Footprint measure permits a simple, straightforward comparison of one group (usually a nation, though the methodology has been applied to sub-national scales, for example, see Wackernagel, 1998) against another group, as well as allowing comparison of the resources required by an economy to the resources available to that economy. The "economy" is treated as one homogenous entity, with the "footprint" of the economy often expressed on a per capita basis (by dividing the total footprint by the population of the economy). When this accounting tool is used to compare resource use between groups, it implicitly ignores any inequalities *within* an economy.

Similarly, the "IPAT" formula, proposed by Ehrlich and Holdren in the early 1970s (Ehrlich & Holdren, 1971) is a frequently used model for conceptualizing environmental impacts resulting from economic activity. The simple formula, $I = PAT$, is used in many introductory texts on Environmental Studies (for example, Miller, 1998, p. 16), and it provides the underlying principle for a vast body of literature on society–environment relationships (Commoner, 1972; Ehrlich & Holdren, 1971; Chertow, 2001; York, Rosa, & Dietz, 2003). According to the IPAT model, a society's environmental "impact" (I) is the product of its "population" (P), multiplied by its "affluence" (A), by its "technology" (T). Population means "overall number of people," while affluence is typically measured in dollars (of GDP or GNP) per capita. The "technology" term, which is designated as "environmental degradation and pollution per unit of resource used," in theory is intended to account for the ecological efficiency of industrial processes. In practice, however, "technology" is rarely measured, and is often used instead to represent "everything else" that population and affluence cannot account for (Chertow, 2001). Importantly, the limited efforts that have been made to put actual numbers into the IPAT equation still rely on a single value for all three factors. Specifically, the model relies on *average* values for both "affluence" and "technology," resulting in the same potential bias as the Ecological Footprint model. For societies where a small fraction of the population consumes many more resources or produces much more pollution than other members, the IPAT model fails to communicate the vast differences in resource consumption that exist within groups.

Even refinements continue to display this difficulty. When the "IPAT" model was revised by York et al. (2003) to "STIRPAT" (STochastic Impacts by Regression on Population, Affluence, and Technology), for example, those authors chose to account for random effects by introducing an "error term." The authors rewrote the model as $I_i = aP_i^b A_i^c T_i^d e_i$, where the coefficients a, b, c, and d can be estimated by regression analysis. Unfortunately, the final estimated model stills relies on average values for affluence and technology, and thus makes no improvement its ability to prevent outliers (such as the fifth U.S. citizen in the simple model presented earlier in this chapter), from influencing what are taken to be typical, normal, or representative values.

One simple step toward resolving the potential bias present in both the Ecological Footprint method of accounting and the IPAT or STIRPAT models would be to use median values instead of means (see discussion in "Getting the Numbers Right" section). While this approach would markedly improve the accuracy of the models in comparing the "typical" impact of one society to the "typical" impact in another society, the models would still focus attention to the middle of the distribution, rather than the "tail," which is where we can expect to get the leverage for pollution reduction efforts. Before changing widely accepted models, and insisting on inspections of resource consumption distributions, it is pertinent to review theoretical perspectives from sociology, in conjunction with mathematical principles, on why we should, or should not, *expect* distributions to be disproportional.

SOCIOLOGY MEETS STATISTICS: SHOULD WE EXPECT DISPROPORTIONALITY?

A key question in the sociological literature, in the late 1970s and 1980s, was whether relationships between the environment and the economy were characterized by "inherent conflict." According to the "core" literature of environmental sociology, economic growth is expected to be associated with greater environmental harm (for a relatively recent assemblage of summaries and further reviews, see the *Handbook of Environmental Sociology* (Dunlap & Michelson, 2002)). By the 1990s, on the other hand, several bodies of work (largely of European origin) began to express nearly the opposite expectations, emphasizing hypothesized *environmental benefits of economic growth.*

One relatively clear example of the "core" literature on *conflict between economic prosperity and environmental protection* is provided by the work of Schnaiberg (1980) and of Schnaiberg and Gould (1994), who argue that there is "an enduring conflict" between environmental protection and economic growth. Their work, focusing on what Schnaiberg has called the

"Treadmill of Production," is based on the argument that economic producers need ever-increasing profits, which they attempt to achieve by means of ever-growing production, leading in turn to ever-increasing environmental impacts. Similar expectations are evident in later work on what O'Connor (1988, 1991) (see also Foster, 1992) has called "the second contradiction of capitalism." In essence, O'Connor sees capitalism as relying on prosperity for legitimatizing economic expansion, creating a need for ever-increasing exploitation of the environment. If the work of Karl Marx (Marx, Moore, Aveling, & Engels, 1889) on "the first contradiction of capitalism" involves the exploitation of workers in the search for producer profit – to the point where the workers are unable to buy the products – O'Connor proclaimed that "the second contradiction" involves such heavy exploitation of natural resources that nature, too, can no longer support the continued survival of capitalism. Under the expectations of the "core" literature, the fundamental conflict between economic activity and environmental welfare *necessitates* environmental destruction, and this *necessity* should explain the observed environmental degradation associated with an economy. Thus, we should expect relatively uniform amounts of "environmental harm per unit of economic activity" (among producers of homogenous products). Specifically, the environmental harm resulting from each producer in the economy is expected to be proportional to its output level (this prediction corresponds to "Case A, within-group Proportionality" in the "Getting the Numbers Right" section of this chapter).

Virtually the opposite expectation – namely that economic growth can benefit the environment – is perhaps most clearly spelled out and explained in a body of sociological work on "ecological modernization," originally proposed in the German language by Huber (1985, 1991), but best-known in English-language form through the work by Spaargaren and Mol (beginning with Spaargaren & Mol, 1992). For purposes of the present chapter, the key argument of work within this tradition is that, in contrast to the relatively pessimistic views in the "core" literature of environmental sociology, environmental problems can best be solved through *further* advancement of technology, including what Spaargaren and Mol (1992) call "super-industrialization." Similar expectations are found in work on "postmaterialism" and on "reflexive modernization." Work by the best-known proponent of postmaterialism argues that prosperity leads to a greater public willingness "to make financial sacrifices for the sake of environmental protection" (Inglehart, 1995, p. 57; see also Abramson, 1997; Brechin & Kempton, 1994, 1997), and scholars within the reflexive modernization school see civil society as becoming a driving force for environmental policymaking in an age of risk. Although the underlying mechanisms are not always spelled out in consistent ways, all of these bodies of work tend to expect the widespread emergence of an "environmental state" (see Frank, Hironaka, & Schofer, 2000; but see also Buttel, 2000; Fisher & Freudenburg, 2001), overseeing the kinds of economic expansion that are beneficial to the environment. It is conceivable, under the "environmental state" theory, that differences in resource efficiencies could lead to unequal distributions of resource consumption per unit of output (whether due to some actors adopting technological advancements, as purported by ecological moder-nization theory, or by certain actors making financial sacrifices, as predicted by postmaterialists). None of these approaches, however, pay significant attention to the environmental implications of differing statistical distribu-tions. In the next section, accordingly, I review statistical properties of two distributions, normal and lognormal, (both of which involve varying degrees of disproportionality) and compare their implications to the expectation of proportionality under the "core" sociological literature.

"Normal" Disproportionality

The probability density function or "distribution" in Fig. 2 is well-known and widely used throughout the social and physical sciences, perhaps

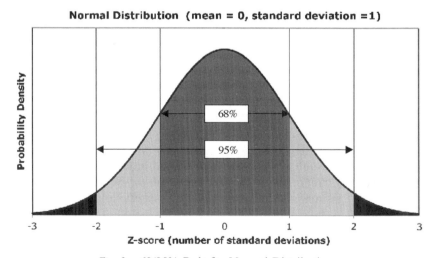

Fig. 2. 68/95% Rule for Normal Distributions.

justifying the name of "Normal Distribution," although it is also called the "Gaussian Distribution" (after Carl Friedrich Gauss) or "bell curve." Many common statistical procedures, such as *t*-tests and Ordinary Least Squares (OLS) regressions, rely on the assumption that data or error terms are normally distributed. As a quick review, a normal distribution theoretically results any time observations are influenced by many small, independent factors that are additive. For example, the distributions of physical properties of individuals in a population (such as height or weight of same sex, adult individuals) are typically well represented by a normal distribution. It seems reasonable to think that the ratio of resource use or pollution production to differing levels of economic activity might be normally distributed within an economy. However, most people would suspect that some sectors of the economy are "inherently" more polluting than others – for example, service-based sectors would be expected to pollute much less than manufacturing sectors. Within a specific industry, resource use or pollution levels among firms might be expected to follow a normal distribution, due to the combined influence of many small differences that exist among firms.

The normal distribution can be completely characterized by two measures – the average and the standard deviation (which can roughly be thought of as the "average distance of the data from the mean").[1] An important aspect of the normal distribution is that it is governed by a simple rule of thumb for predicting the degree and abundance of extreme observations, known as the 68/95% rule (see Fig. 2). This rule says that 68% of the data will be contained within one standard deviation of the mean (note: the mean is equal to the median for a normal distribution). The complement to this statement is that 32% of the data will be more than one standard deviation above or below the average. Since the distribution is symmetric, we can expect 16% of the observations to be more than one standard deviation above the mean, and 16% to be more than one standard deviation below. Similarly, the rule says that 95% of the observations will have values within two standard deviations of the average, meaning that 2.5% will have values more than two standard deviations above or below the arithmetic mean.

An important implication of the simple 68/95% rule is that we should *expect* a number of polluters to emit a disproportionate amount of pollution per unit of output (which will be referred to as the "resource efficiency" or "emissions/unit output"), when resource efficiency is normally distributed. In the case of a group of firms producing homogeneous products, the 68/95% rule would tell us that we could expect 16% of firms to emit an amount of pollution per unit of production that is more than one standard deviation above the mean. As an illustrative example, consider 100 firms in a specific industry, such as Primary Nonferrous Metal producers, within the United States. We could expect 16 of them to be polluting at levels beyond one standard deviation above normal, while about 2 of those 16 firms could be expected to pollute at levels that were more than two standard deviations above the industry average. Even if the firms were simply to be normally distributed, in other words – contrary to the way in which within-group pollution has treated to date within the "core" literature of sociology – all polluters would *not* be interchangeable. Instead, if pollution were to be normally distributed, we should *expect* to find some outliers who pollute at levels well beyond the average, while an equivalent number of firms would be equally far below the average in their pollution efficiency.

If we believe that pollution efficiency is normally distributed among firms, rather than being directly proportionate to levels of output, a pollution prevention policy that targets the "extreme polluters" could potentially decrease emissions substantially, at relatively low regulatory costs, compared to a pollution prevention policy that focused on industry-wide decreases in emissions, or one which relied on voluntary actions of polluters (for specific example, see Nowak et al., 2006, p. 169). Still, in this kind of a "normal" distribution, the ultra-efficient firms would counterbalance the highly polluting firms, meaning that the arithmetic mean of emissions/unit output would be regarded by most statisticians as a suitable measure of "central tendencies," or typical values, for the overall distribution. This property, on the other hand, is not true of skewed distributions, as is discussed below.

"Log-Normal" Disproportionality

An even more extreme form of within-group disproportionality exists when observations are log-normally distributed. The log-normal distribution has a certain resemblance to the normal distribution, but is positively skewed, as shown in Fig. 3, below. Limpert, Stahel, and Abbt (2001, p. 351) argue "… the reasons governing frequency distributions in nature usually favor the log-normal distribution, whereas people are in favor of the normal." These authors have noted any number of examples from physics, chemistry, medicine, and environmental science for which the log-normal distribution fits better than the normal distribution.

A log-normal distribution theoretically results when observations are influenced by many small, independent factors (as in a normal distribution),

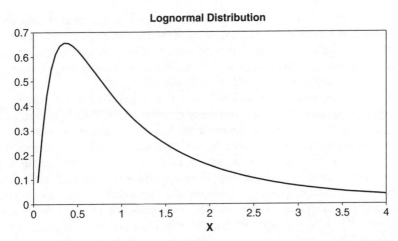

Fig. 3. Log-Normal Distribution with Median Equal to 1, and the Multiplicative Standard Deviation of *e* (2.718).

but where the "typical" observation has a low value, the standard deviation is high (relative to the mean), and the values cannot be negative, or when independent factors combine in a multiplicative fashion (rather than additively, as in a normal distribution). The log-normal distribution looks similar to a normal distribution, but the left-hand tail is truncated (it is also called a "left-censored" distribution). In terms of the distribution of pollution efficiency, values of emissions-to-output ratios can only be "non-negative," and a small but growing number of authors have presented evidence that it is reasonable to expect the typical polluter within groups to be relatively efficient, and to expect a considerable amount of variation among firms within an industry (see Freudenburg, 2005, 2006; Nowak et al., 2006).

What are the implications of a log-normal distribution? The name "log-normal" reflects the fact that taking the natural logarithm of the observations will result in a normal distribution. Since taking the logarithm "undoes" exponentiation, log-normal distributions tend to have some observations that are orders of magnitude beyond the "typical value" (which is characterized by the median, not the mean). This fact (coupled with the absence of negative values) leads to a long tail on the right-hand side of the distribution, and a large number of observations "piled up" on the left-hand side, as pictured in Fig. 3.[2] To consider emissions among polluters, let us assume a log-normal distribution of 100 polluters whose median pollution efficiency is 7.4 emissions/unit output, with a multiplicative standard deviation of 2.2 emissions/unit output. Using these hypothetical values, we find that 65% of polluters actually emit less than "average" (the mean is 10 emissions/unit output). Additionally, we can expect the top two firms to emit at least 36 emissions/unit output – almost five times as much as the "typical" polluters.

This simple discussion shows that it is worth asking whether the production of pollution is log-normally distributed; in situations where this is the case, the outliers, or highly polluting firms, could be expected to release pollutants at levels much higher than those typical of the industry. Since the average itself is highly influenced by extremes, models or assumptions that focus on the "average (mean) polluter" within an industry, or the "per-capita pollution" for the industry, will *overestimate* the central, or typical value for the vast majority of firms within the industry. Thus, it may appear that there is more pollution associated with "average" firms within an industry than would be the case if analyses were to focus on the "typical" or median firm – and our analyses may lead to policy responses (such as treating all firms as "equals") that fail to reflect the actual distribution of problems. This phenomenon may explain why polluters often argue that pollution cleanup costs are "too costly." Focusing on the "average," rather than typical, polluter could make "cleaning up an industry" appear much more difficult than it actually may be. In the following section, I review several case studies of the within-group distribution of pollution production.

BEYOND THEORY: EMPIRICAL EVIDENCE OF DISPROPORTIONALITY IN POLLUTION PRODUCTION

The United Nations Development Programme (UNDP, 1998) reports:
    … Inequalities in consumption are stark. Globally, the 20% of the world's people in the highest-income countries account for 86% of total private consumption expenditures – the poorest 20% a minuscule 1.3%. More specifically, the richest fifth:

- Consume 45% of all meat and fish, the poorest fifth 5%
- Consume 58% of total energy, the poorest fifth less than 4%
- Have 74% of all telephone lines, the poorest fifth 1.5%
- Consume 84% of all paper, the poorest fifth 1.1%
- Own 87% of the world's vehicle fleet, the poorest fifth less than 1%

These well-known statistics are part of a large and growing body of work on "environmental justice," (summarized elsewhere in this volume by Bullard, in press; Mohai, in press; Taylor, in press; see also Boyce in press; Harlan et al., in press), which explicitly addresses the potential inequities in the environmental impacts *experienced by* different social groups. Only recently, however, have sociologists and physical scientists began to explore potential widespread inequities in environmental impacts *created* by *social* factors within groups.

In the 1990s, interdisciplinary work began on a pair of projects supported by the National Science Foundation (NSF) at the University of Wisconsin-Madison (one a Long-Term Ecological Research (LTER) Project and the other a newer program of Integrative Graduate Education and Research Training (IGERT)) focusing on "Lakes and Society." The projects brought together a number of researchers, including Steve Carpenter, Bill Freudenburg, and Pete Nowak, the latter two of whom would go on to lay the foundation for the concept of "disproportionality" as a pattern that bridges the physical and social sciences. Steve Carpenter, the lead PI on the LTER, is an ecologist who was interested in species-abundance trends, among other phenomena. As he discussed with Nowak and Freudenburg, studies of species-abundance patterns commonly found a log-normal pattern, where many species occurred with low abundance, and very few species occurred with high levels of abundance.

Professor Freudenburg (1997) had by that time begun to argue that sociological work on the environment should pay greater attention to the firms putting out the highest levels of pollution, but discussions with Professor Carpenter led both Professors Freudenburg and Nowak to discuss the extent to which the notion of log-normal distributions might provide a useful framework for describing the patterns that had been emerging in their studies of environment–society relationships.

Both Freudenburg and Nowak had by that time reached the conclusion that the broader "economy vs. environment" debate (see "Sociology meets Statistics") was in effect *too* broad. In their view, and in the view of several social scientists working with them on the NSF projects (see especially Fisher & Freudenburg, 2001, 2004; Nowak et al., 2006), the available empirical record included a number of cases that appeared to be at least reasonably consistent with the expectations being expressed on both sides of the broader, or macrosociological, divide. The irony, of course, is that if both "sides" have offered empirical evidence for their own perspectives, then neither side could be completely or consistently correct. Freudenburg and Nowak had thus both become increasingly interested in examining specific cases of environmental harm that might throw light on *the extent to which* any given case of environmental harm might actually be "necessary" for (and therefore proportionate to) economic output, providing insights on *the conditions under which* one point of view or the other might be likely to be more accurate.

In his own research on toxic releases, in particular, Freudenburg (1997) had noticed that certain sectors of the economy produced dramatically more toxic releases than others, whether the releases were measured in terms of raw pounds or in terms of toxicity. Even within a specific industry or sector of the economy, certain firms had emissions levels well beyond the "average" level for the industry or sector in question – and those differences were evident even after accounting for the size of the facilities, in terms of either number of employees or output of products.

Freudenburg reports that he had previously noted that the levels of pollution within specific industries tended to look vaguely log-normal in their distributions, but that it was the casual conversation with Steve Carpenter, followed by more intensive conversations with Pete Nowak, that helped to crystallize his thinking. In a series of papers, Freudenburg and Nowak (2000) (see also Freudenburg, 1997, 2005, 2006; Nowak & Cabot, 2004; Nowak et al., 2006) postulated that "disproportionality" between economic activity and environmental harms (e.g., toxic releases) needed to be examined as a variable that could be used to move the social science literature on environment–society relationships beyond broad or black/white assertions. Examining actual levels of disproportionality, to be more specific, could allow researchers to measure the extent to which economies (or specific economic industries or sectors) could *appear* to be at odds with the environment (as predicted by the "core" works of environmental sociology) – at least when measured by mean or aggregate levels of pollution – even though the *majority* of firms in that industry or sector might support the idea of compatibility between the economy and the environment, being responsible for much lower levels of environmental degradation than the mathematical average for their own industries or sectors.

To quantify the levels of disproportionality in toxic releases, Freudenburg (1997) began with the Environmental Protection Agency's (EPA's) Toxic Release Inventory for the year of 1993 – one of the first years to become widely available to researchers in CD-ROM format. In an effort to communicate his results to the research community, Freudenburg (2005) ultimately utilized the Gini coefficient, a measure of inequality that has been widely used to describe the degree of inequality in a society's wealth, or incomes. Values for Gini coefficients range from zero to one, with zero describing a society having perfect equality – one in which incomes (or pollution) are spread equally among all persons (or facilities) – and a value of one describing a society of perfect inequality, in which one person (or facility) is associated with 100% of the income (or pollution), and all other persons (or facilities) have no income (or pollution). As noted by Freudenburg (2005, p. 96),

In practice, coefficient values range from around .2 for historically equalitarian countries such as Bulgaria or Hungary to around .6 for nations where powerful elites dominate the economy, with the world's highest coefficient today being associated with Sierra Leone, at .62. Most present-day European countries and Japan range from around .25–.32, while most African and South American countries – and in recent years, the United States – have had Gini coefficients in the range of .45–.50

Using the same coefficient for analyzing inequalities in levels of toxic emissions, Freudenburg found that, even when data were compared only across the seven industries that made up the most toxic industrial sector of the U.S. economy – Standard Industrial Classification (SIC) category 33, Primary Metals – the disproportionality levels were higher than in any national study he had ever encountered. South Africa, during the era of Apartheid, had been found to have a Gini coefficient of 0.71, but the Gini coefficients for disproportionality of emissions were consistently higher, even after controlling, statistically, for the differing size of one industry vs. another. After controls were imposed for the sizes of payrolls and for the number of employees, "the Gini coefficients actually became even more extreme, rising to 0.817 and 0.821, respectively." Sharpening the focus ever further, and comparing the 62 enterprises or facilities within the most toxic sector (SIC 333, or Primary Nonferrous Metals), Freudenburg (2005, p. 100) found that

Rather than becoming more even, the results become even more disproportionate, leading to a higher Gini coefficient than this author, at least, has ever encountered in any other context: .975. So disproportionate are the emissions from this sector, that a single facility – Magnesium Corporation of America, in Rowley, Utah – accounted for more than 95 percent of the toxicity emitted from the entire 333 SIC code, or for that matter, roughly 75 percent of the toxicity associated with the riskiest two– digit sector of the entire economy. Although this facility may indeed be an "outlier" in many respects, in short, it produces such a high level of toxic emissions that this single facility has more influence on the overall levels of toxic emissions from this sector of the economy than do all other facilities in the same sector, even in combination.

The firm size and output could not be controlled statistically at the facility level of comparison, due to lack of data availability, but rather than seeing this facility as being an unusual case that should simply be ignored, Freudenburg began to describe such facilities as being better understood as examples of what another paper (Berry, Freudenburg, & Howell, 2004) called "the tail that wags the distribution." Still, to investigate whether or not the Magnesium Corporation of America might be an anomaly, Freudenburg recalculated the Gini coefficient for the remaining 61 facilities in the industry, finding that, as would be expected, the resultant 61-facility comparison did show a decline in the Gini coefficient, although that decline was "only back to the same general range seen in the earlier Gini coefficients of this chapter, or 0.735 – a coefficient that still remains above the levels of income inequality associated with South Africa during the time of apartheid, as well as being substantially higher than the coefficient for income inequality in any nation of the world today."

Freudenburg's research, although exploratory in nature, has presented convincing evidence that remarkably high degrees of disproportionality have gone unnoticed, and that focusing on aggregate measures of environmental performance (such as averages or totals) misses both the disproportionate polluters that create a substantial fraction of all damage *and* the majority of firms, which are polluting less than the average or mean.

The work by Freudenburg's colleague in Madison, Pete Nowak, has developed complementary findings while working on a very different kind of pollution problem. In research that tied in with his collaborations with Professors Carpenter and Freudenburg, Nowak has long been studying phosphorus (P) loading within the Pheasant Branch Creek Watershed of Lake Mendota, near Madison, Wisconsin. His research addressed the question: "How could land use and management have changed significantly in some regions of the Lake Mendota watershed over the prior two decades without prompting a decrease in P [phosphorus] loads to the lake?"

Nowak observed patterns of disproportionality in the relative impact of polluters, at various scales of analysis, which were strikingly similar to Freudenburg's findings. In the case of Lake Mendota, however, Nowak has found that the key determinants of the impact of human actions on the ecosystem involved the biophysical properties of the environment, *in combination with* inappropriate social behaviors. Nowak's key findings involved what he and his subsequent colleagues have called "coarse-scale" and "fine-scale" disproportionality, where "coarse-scale" disproportionality involved long-term spatial and temporal differences in P loading in the Pheasant Branch subwatershed, and fine-scale disproportionality involved phosphorous levels within individual agricultural fields, as measured in a shorter time period.

The important point is that these biophysical features (i.e., excessive soil P levels) were not distributed uniformly across the landscape in the Pheasant Branch subwatershed. Instead, the "hot spots" of soil P were spatially distributed in a pattern that coincided with or were proximate to former or current livestock operations. Watershed P budgets' (Bennett et al., 1999) and P dynamics for the entire Lake Mendota system (Reed-Anderson et al., 2000), however, were based on average P levels across the entire watershed. While both studies showed an average buildup of P in agricultural soils across time, neither addressed the heterogeneous spatial patterns of soil P induced by human behaviors ... (Nowak et al., 2006, p. 161)

Nine of the 10 commercial farms in the watershed study area allowed researchers on their fields to obtain soil samples based on a 1 ha sampling grid (Cabot & Nowak, 2005) ... . The majority of the soil test results are in the high or excessively high range for P values, with a clear sub-set of outliers that have values up to 900% above the sample mean. (Nowak et al., 2006, p. 166)

From these findings, Nowak et al. take issue with the tendency to describe the outliers as "bad actors," concluding instead that "This disproportionate outcome occurs, not because the behavior of the minority is especially egregious or deviant, but because their actions are inappropriate behaviors taking place in biophysically vulnerable settings or time. For example, the all too common term of "bad actor" is only partially correct; both the "acting" and the "stage" for that action need to be used in forming such ill-advised value judgments." (Nowak et al., 2006, p. 158).

Nowak and his colleagues have effectively combined the disproportionality among social actors within a group (as found in Freudenburg's research on toxic releases) with inequities known to exist in biophysical phenomema (as found in biological work on species abundance curves) and processes (such as nutrient export and soil loss). His analysis highlights a type of reasoning called "systems thinking," which postulates that a system cannot be understood by reducing it to (and understanding only) its parts. Instead, understanding synergies between social actions and biophysical settings is necessary to predict and explain environmental outcomes.

The pioneering work of Carpenter, Freudenburg, and Nowak suggests that the patterns of disproportionality in the *creation* of environmental damage have largely gone unnoticed and/or taken for granted in past analyses of people and the environment. Their early results show that tangible, and often very serious outcomes, such as toxic pollution and degradation of land and waters, can be highly influenced by a few "outliers." Outliers, by definition, are associated with values far away from the norm, but just how far from "typical" are they? When polluters are not interchangeable, it is pertinent to question how much of the pollution is a result of small number of heavy polluters, which is the topic of the next section.

THE TRAGEDY OF THE COMMONS: HOW TO MANAGE A NOT-SO-"COMMON" PROBLEM

Hardin's classic (1968) *Science* article, "The Tragedy of the Commons," paints a portrait of a society that allows unrestricted access to a common (shared), yet finite pool of resources (in Hardin's example, the limitation is the biological carrying capacity of the commons). In Hardin's imagined society, based loosely on the English grazing "commons," each profit-seeking individual will have incentive to exploit more and more of the pasture. Doing so imposes an external cost on other grazers, and ultimately degrades the commons. Hardin uses the example of cattle grazing on a pasture, where the pasture is the shared resource, and each individual derives private benefit from adding another cow to the pasture. The "cost" associated with an additional cow's grazing (once the carrying capacity of the system is exceeded) may come in the form of depletion of grass, soil compaction and subsequent loss, and removal of nutrients, but these costs are spread among all of the society's pasture-sharing members. Thus, if a person adds another cow to the pasture, s/he derives sole benefit from the addition, but only pays a fraction of the cost.

To Hardin, and many people living in the present-day U.S., it is intuitive that each member of the group, trying to maximize his or her welfare, will add as many cows as s/he can afford to add. If we imagine a pasture that can support only 90 cows each year, and 10 extra cows are added by "economic welfare-maximizing" individuals, the pressure from the cows added beyond the pasture's carrying capacity will degrade the pasture. Therein lies the "tragedy." The exploitation of common-pool resources is "inevitable," due to the payoffs faced by each individual. Importantly, at least in Hardin's tale, each member in the society can be seen as interchangeable, in that they all use the same decision-making logic.

Under the circumstances outlined above, a "rational manager" (or historically, the villagers who owned the cattle) would see an obvious solution of removing the 10 extra cows. The impact, under this simple example, clearly is "proportionate" to the number of cows involved, so removing 10 percent of the cows (10 out of the 100) could reduce the impact by 10 percent, bringing the number of grazers back to the carrying capacity of the pasture. Importantly, the cows are "interchangeable," it does not matter *which* cows are removed, only how many, since each cow consumes the same fraction of the pasture's carrying capacity.

In the case of toxic emissions in the U.S., by contrast, Freudenburg, Berry, and Howell (2006) found that the most heavily polluting mines or primary metals facilities did not even need to be "removed" from the economy. If instead the highest 10 percent of polluting facilities within the primary non-ferrous metals industry, for example (4 facilities out of 37 facilities in the industry for which economic data were available at the 3-digit SIC level), were simply to have reduced their emissions to sales ratio to the median level of other facilities in the same industry, the net result would have been a 83.5% reduction in the total toxic emissions for the entire industry. Even if only the *single* highest-polluting facility were merely to reduce its emissions *per dollar* to the median pollution level for that industry, over half of that industry's entire total of toxic emissions (58.2%) would be avoided.

Clearly, these preliminary findings indicate that pollution production should not be assumed to be a "Tragedy of the Commons." In this case, instead, high amounts of pollution (per unit of economic production) were actually quite uncommon. The type of scenario building described above (where the pollution efficiency of the highest polluting firms within an industry are decreased to the median level of the group) may offer a useful starting point for understanding how much of a difference the dispropor-tionate polluters make.

Research on disproportionality is still at a sufficiently early stage that it is not possible at present to say whether such reductions are higher, lower, or about average for the kinds of environmental improvements that could be obtained in other sectors of the economy. Still, even if the case of this one sector of the economy proves to be relatively extreme, it is clear that focusing more attention on "outliers" can provide opportunities to reduce pollution at a lower marginal cost than policies that are focused on the "average" polluter.

Under conditions of disproportionality, simply blaming the pollution problems of an industrialized economy on the "average" individual is not only inaccurate, but is badly misleading. In fact, using the average not only misses the "outliers" – the firms that are responsible for the lion's share of the harm – but it also distracts attention from the vast majority of actors who are able to produce goods at pollution levels significantly below the mean. Future analyses need to pay attention not only to "how much" pollution, but also to "who" is polluting "how much." From a pollution policy and management perspective, it is imperative to be able to detect disporportionality.

DETECTING DISPROPORTIONALITY: GOODNESS-OF-FIT STATISTICAL TESTS

Future research on disproportionality will be most effective if the detecting of log-normal distributions – those that are characterized by a few outliers that skew the mathematical average – is "operationalized." One promising option for moving in that direction, is the use of statistical "goodness-of-fit" tests, which allow the researcher to test whether the data come from various distributions (such as normal or lognormal ones).

A useful starting point for this topic involves the fact that the expectations of the various sociological theories of society–environment interactions (reviewed in the "Sociology meets Statistics" section of this paper) can be tested using statistical goodness-of-fit tests. Under the expectations of the "core literature," which predicts a fundamental conflict between economic activity and environmental welfare, environmental harm should be proportional to economic activity. If we allow for small and unrelated random effects to contribute additively to the amount of toxic releases per unit output among firms in an industry, then the probability distribution might be expected instead to be roughly normal (Gaussian distributed). Finally, the growing body of new evidence, suggesting the existence of considerable disproportionality, would cause us to expect the probability distribution of environmental harm per unit of economic activity to be positively skewed, with a small number of economic actors accounting for the largest inputs of environmental pollution per unit of economic activity. Table 2 summarizes the distribution that would be expected under each hypothesis.

In general, all goodness-of-fit (GOF) tests use the same null hypothesis – namely, that the data are sampled from a specified distribution. The test statistics report the probability of findings that deviate as much, or more, from the observed sample data, if the null hypothesis were to be true. The tests differ in how they quantify the deviation of the observed sample distribution from a specified distribution. Preliminary results (Berry, Freudenburg, & Howell, 2005) indicate that, for eight different ways of examining toxic emissions (at the 3-digit SIC code level of analysis, controlling for economic output), the normal probability distribution was rejected (at a 0.0001 significance level) in every case. By contrast, in seven

Table 2.   Expected Distributions Under Various Sociological Theories. The Author Recommends Using the D'Agostino–Pearson Goodness-of-fit Test to Characterize Distributions (D' Agostino, Berlanger, & D'Agostino, 1990).

Hypothesized Relationship between Environmental Harm and Economic Activity	Sociological Theory	Expected Probability Distribution
Proportional (linear)	"core" literature	None: one value characterizes all polluters (pollution necessary to produce good/unit output)
Proportional + random effects	"core" literature, relaxed (largely unexamined in literature to date)	Normal
Disproportional, positively skewed	Largely unexamined, but expected in disproportionality literature and potentially compatible with ecological modernization	Lognormal

out of the eight cases, the log-normal probability distribution could not be rejected. Interestingly, in the one case where the lognormal distribution was rejected, it appeared that the data was skewed too far to the right – that is, the highest polluting facilities were even more extreme than would be expected even from lognormal distribution.

GOF tests provide a straightforward method of detecting disproportionality, but, more importantly, examining the distribution of pollution production permits the researcher to notice "who" is responsible for "how much" of the pollution.

CONCLUSION

To portray society–environment relationships accurately, and thus to enable policy makers to select effective measures for reducing pollution and environmental harm, there is an urgent need to understand the distribution of within-group resource use and pollution production. The case studies discussed in this review are still limited in number, but they tend to support the Disproportionality Hypothesis: emissions levels within a specific economic industry tend not to be proportional to the level of economic output, due to a small number of firms which pollute much more than their "fair share." Collectively, these disproportionate polluters represent an opportunity for high rates of return, in the form of pollution reduction, if policies target the outliers.

If this pattern of findings continues to be found in future research, it would mean that the widespread "Proportionality" hypothesis – the assumption that underlies most between-group comparisons of resource use and pollution production – significantly overstates the amount of pollution that is inherently associated with a given level of industrial output. It would also mean that the amount of pollution that is actually "necessary" for any given level of industrial output might be substantially lower than is the level predicted by "per capita" approaches. In cases where within-group pollution production is log-normally distributed, similarly, regulations that target highly polluting firms will likely reduce the overall pollution for an industry at a much lower cost than can be achieved by regulations that require incremental reductions from all firms.

NOTES

1. A higher standard deviation corresponds to observations that are more variable, or further spread apart from the expected value.

2. The degree to which the distribution is positively skewed depends on how large the spread of the data (measured by the standard deviation) is relative to the value of the median.

REFERENCES

Abramson, P. R. (1997). Postmaterialism and environmentalism: A comment on an analysis and a reappraisal. *Social Science Quarterly, 78,* 21–23.

Berry, L., Freudenburg, W. R., & Howell, F. M. (2004). The tale is in the tail: Assessing the disproportionality of toxic releases. Paper presented at the 10th international symposium on Society and Resource Management, Keystone, CO, June.

Berry, L., Freudenburg, W. R., & Howell, F. M. (2005). Studying the tail that wags the distribution: Testing the hypothesis of disproportionality against assumptions of proportionate access. Paper presented at the 68th annual meeting of the Rural Sociological Society, Tampa, FL, August.

Boyce, J. K. (in press). Is inequality bad for the environment? *Research in Social Problems and Public Policy* (Vol. 15).

Brechin, S. R., & Kempton, W. (1994). Global environmentalism: A challenge to the postmaterialism thesis? *Social Science Quarterly, 75,* 245–269.

Brechin, S. R., & Kempton, W. (1997). Beyond postmaterialist values: National versus individual explanations of global environmentalism. *Social Science Quarterly, 78*(1), 16–25.

Bullard, R. (in press). Equity, unnatural man-made disasters, and race: Why environmental justice matters. *Research in Social Problems and Public Policy* (Vol. 15).

Buttel, F. H. (2000). World society, the nation-state, and environmental protection. *American Sociological Review, 65*(1), 117–121.

Cabot, P., & Nowak, P. J. (2005). Planned versus actual outcomes as a result of animal feeding operation decisions for managing phosphorus. *Journal of Environmental Quality, 34*(3), 761–773.

Chertow, M. (2001). The IPAT equation and its variants: Changing views of technology and environmental impact. *Journal of Industrial Ecology, 41*, 3–29.

Commoner, B. (1972). The environmental cost of economic growth. In: R. G. Ridker (Ed.), *Population, resources and the environment* (pp. 339–363). Washington, DC: Government Printing Office.

Dunlap, R. E., & Michelson, W. (Eds). (2002). *Handbook of environmental sociology.* Westport, CT: Greenwood Press.

D'Agostino, R. B., Berlanger, A., & D'Agostino, R. B. (1990). A suggestion for using powerful and informative tests of normality. *The American Statistician, 44*(4), 316–321.

Ehrlich, P., & Holdren, J. (1971). Impact of population growth. *Science, 171*, 1212–1217.

Fisher, D., & Freudenburg, W. R. (2001). Ecological modernization and its critics: Assessing the past and looking toward the future. *Society and Natural Resources, 14*, 701–709.

Fisher, D., & Freudenburg, W. R. (2004). Post-industrialization and environmental quality: An empirical analysis of the environmental state. *Social Forces, 83*(1), 157–188.

Flavin, C., & Gardner, G. (2006). China, India and new world order. In: L. Starke (Ed.), *State of the world, 2006* (pp. 3–23). New York: W. W. Norton & Company (for Worldwatch Institute).

Foster, J. B. (1992). The absolute general law of environmental degradation under capitalism. *Capitalism, Nature, Socialism, 2*(3), 77–82.

Frank, D. J., Hironaka, A., & Schofer, E. (2000). The nation-state and the natural environment over the twentieth century. *American Sociological Review, 65*(1), 96–117.

Freudenburg, W. R. (1997). The double diversion: Toward a socially structured theory of resources and discourses. Paper presented at the annual meeting of American Sociological Association, Toronto, Ontario, August.

Freudenburg, W. R. (2005). Privileged access, privileged accounts: Toward a socially structured theory of resources and discourses. *Social Forces, 94*(1), 89–114.

Freudenburg, W. R. (2006). Environmental degradation, disproportionality, and the double diversion: The importance of reaching out, reaching ahead, and reaching beyond. *Rural Sociology, 71*(1), 3–32.

Freudenburg, W. R., Berry, L., & Howell, F.M. (2006). Disproportionality and distraction in research and theory on environment-society relationships. Paper presented at the 12th international symposium on Society and Resource Management, Vancouver, BC, Canada, June.

Freudenburg, W. R., & Nowak, P. J. (2000). Disproportionality and disciplinary blinders: Understanding the tail that wags the dog. Paper presented at the 8th international symposium on Society and Resource Management, Bellingham, WA, June.

Guralnik, D. (Ed.). 1986. Average. Webster's new world dictionary of the American Language (p. 96, 2nd college ed.). New York: Simon and Shuster.

Hardin, G. (1968). The tragedy of the commons. *Science, 162*, 1243–1248.

Harlan, S. L., Brazel, A. J., Jenerette, G. D., Jones, N. S., Larsen, L., Prashad, L., & Stefano, W. L. (in press). In the shade of affluence: The inequitable distribution of the urban heat island. *Research in social problems and public policy* (Vol. 15).

Harrison, P., & Pearce, F. (2001). In: Victoria Dompka Markham (Ed.). AAAS atlas of population and environment. American Association for Advancement of Science, University California Press.

Huber, J. (1985). *Die Regenbogengesellschaft. Okologie und Sozialpolitik.* Frankfurt am Main: Fisher Verlag.

Huber, J. (1991). Unternehmen Umwelt. Weichenstellungen fu Eine okologische Marktwirtschart.

Inglehart, R. (1995). Public support for environmental protection: Objective problems and subjective values in 43 societies. *PS: Political Science and Politics, 28*, 57–72.

Limpert, E., Stahel, W. A., & Abbt, M. (2001). Log-normal distributions across the sciences: Keys and clues. *BioScience, 51*(5), 341–352.

Marcotullio, P., & McGranahan, G. (2005). Urban Systems, Chapter 27. In: J. Eades, E. Ezcurra & A. Whyte (Eds), *Millennium ecosystem assessment, ecosystems and human well-being, Volume 1: Current state and trends* (pp. 795–826). Washington, DC: Island Press.

Marx, K., Moore, S., Aveling, E. B., & Engels, F. (1889). *Capital: A critical analysis of capitalist production.* New York: Appleton & Co.

Miller, G.T. (1998). *Living in the environment: Principles, connections, and solutions* (10th ed., p. 21). Belmont, CA: Wadsworth Publishing Company.

Mohai, P. (in press). Equity and the environmental justice debate. *Research in Social Problems and Public Policy* (Vol. 15).

Nowak, P. J., & Cabot, P. E. (2004). The human dimension of resource management programs. *Journal of Soil and Water Conservation, 59*(6), 123A–135A.

Nowak, P. J., Bowen, S., & Cabot, P. E. (2006). Disproportionality as a framework for linking social and biophysical systems. *Society and Natural Resources, 19*, 153–173.

O'Connor, J. R. (1988). Capitalism, nature, socialism: A theoretical introduction. *Capitalism, Nature, Socialism, 1*(1), 11–38.

O'Connor, J. R. (1991). On the two contradictions of capitalism. *Capitalism, Nature, Socialism, 2*(3), 107–109.

Schnaiberg, A. (1980). *The environment: From surplus to scarcity.* New York: Oxford University Press.

Schnaiberg, A., & Gould, K. A. (1994). *Environment and society: The enduring conflict.* New York: St. Martin's Press.

Spaargaren, G., & Mol, A. P. J. (1992). Sociology, environment, and modernity: Ecological modernization as a theory of social change. *Society and Natural Resources, 5*, 323–344.

Taylor, D. (in press). Diversity and the environment: Myth-making and the status of minorities in the field. *Research in Social Problems and Public Policy* (Vol. 15).

United Nations Development Programme (1998). Overview: Changing today's consumption patterns for tomorrow's human development. Human Development Report, 1998, 1–17. Retrieved on April 20, 2007, from http://hdr.undp.org/reports/global/1998/en/

Wackernagel, M. (1998). The ecological footprint of Santiago de Chile. *Local Environment, 3*(1), 7–25.

Wackernagel, M., & Rees, W. (1996). *Our ecological footprint: Reducing human impact on the earth.* New Society: Gabriola Island, Canada.

Wackernagel, M., & Young, D. J. (1998). The ecological footprint: An indicator of progress toward regional sustainability. *Environmental Monitoring and Assessment, 51*(1–2), 511–529.

World Wildlife Fund. (2006). In: C. Hails (Ed.), *Living Planet Report, 2006.* Retrieved on April 25, 2007, from http://www.panda.org/news_facts/publications/living_planet_report/lp_2006/index.cfm

York, R., Rosa, E. A., & Dietz, T. (2003). Footprints on the earth: The environmental consequences of modernity. *American Sociological Review, 68*(2), 279–300.

IS INEQUALITY BAD FOR THE ENVIRONMENT?

James K. Boyce

ABSTRACT

By respecting nature's limits and investing in nature's wealth, we can protect and enhance the environment's ability to sustain human well-being. But how humans interact with nature is intimately tied to how we interact with each other. Those who are relatively powerful and wealthy typically gain disproportionate benefits from the economic activities that degrade the environment, while those who are relatively powerless and poor typically bear disproportionate costs. All else equal, wider political and economic inequalities tend to result in higher levels of environmental harm. For this reason, efforts to safeguard the natural environment must go hand-in-hand with efforts to achieve more equitable distributions of power and wealth in human societies. Globalization – the growing integration of markets and governance worldwide – today poses new challenges and new opportunities for both of these goals.

PROLOGUE

In the mid-1970s, I lived in a rural village in northwestern Bangladesh, in one of the poorest parts of a poor country. Bangladesh had just had a

Equity and the Environment
Research in Social Problems and Public Policy, Volume 15, 267–288
ISSN: 0196-1152/doi:10.1016/S0196-1152(07)15008-0

famine in which some 200,000 people perished. The famine was caused not by an absolute shortage of rice, the staple food of the population, but rather by a combination of grain hoarding by merchants and government ineptitude and corruption. The village where I lived was located in the most famine-stricken district of the country.[1]

To the eyes of a young American, a striking feature of Bangladeshi village life – apart from the poverty of the people – was the virtual absence of negative environmental impacts from human activities. The villagers farmed rice and jute much as their ancestors had for centuries. Agrochemicals had only begun to appear on the scene, and village farmers used them sparingly, if at all. Across the country, Bangladeshi farmers grew some 10,000 different varieties of rice adapted to microclimatic variations in rainfall, flood depths, temperature, and soil type, making the country a storehouse for genetic diversity of humankind's most important food crop. Hundreds of fish species – more than in all Europe – lived in the country's rivers, ponds, and rice paddies, supplying most of the animal protein in the Bangladeshi diet.

There was no trash-collection service in the village, and no need for one, for the villagers produced little, if any, solid waste. Practically everything they consumed – food, cooking fuel, housing materials, herbal medicines – was harvested from their local environment. Crop residues and manures were returned to the earth or burned as fuel. Metal items were carefully recycled, tin containers fashioned into building supplies. Few villagers had ever seen plastic: local children amused themselves by repeatedly dropping a red plastic cup I had brought with me to the village, showing their friends that it did not break. Despite their poverty – in some ways, because of it – they did not harm the environment, on which their livelihoods depended.

One time, while visiting Dhaka, Bangladesh's capital city, I stumbled upon what may have been the country's first environmental campaign. The government – a one-party state headed by a once-popular politician who recently had declared himself "president for life" – had just announced a campaign of "urban beautification." Dhaka's sprawling slums, which had multiplied during the famine as starving people from the countryside migrated to the city in search of work or relief, were razed to the ground. Their inhabitants were brusquely herded onto trucks that deposited them outside town, far from the eyes and consciousness of the city's middle- and upper-class residents.

The human costs of this policy were vividly brought home to me by a scene I witnessed in front of Dhaka's general post office. An emaciated woman and her baby were sitting on a dirty cloth spread on the sidewalk.

Passers-by occasionally dropped a coin. When I emerged from the post office a few minutes later, the woman was gone, perhaps trailing after a well-to-do stranger to plead for alms. Then a police truck drove by, its bed full of destitute people being relocated. Spying the baby, the truck stopped and a policeman dismounted. He unceremoniously tossed her into the back of the truck, which lumbered off in search of more human cargo. I do not know whether they picked up the mother.

The irony was inescapable and terrible: In a land where they lived lightly on the earth, the poor themselves were regarded as pollution.

WHAT IS ENVIRONMENTAL HARM?

What does it mean to say something is "bad for the environment" or "good for the environment"? These value judgments rest, implicitly or explicitly, on ethical criteria by which we distinguish better from worse.

A criterion that has gained many adherents in the past two decades is "sustainable development." This was defined by the World Commission on Environment and Development in its 1987 manifesto, *Our Common Future* (also known as the Brundtland Report, after commission chair Gro Brundtland), as development that "meets the needs of the present without compromising the ability of future generations to meet their own needs" (World Commission on Environment and Development, 1987). By this criterion, "environmental harm" means actions that compromise the ability of future generations to meet their needs. Conversely, "environmental improvements" would refer to actions that enhance the ability of future generations to meet their needs.

The Brundtland criterion has the merits of affirming the importance of human well-being and our responsibility to future generations. But as Nobel Prize-winning economist Amartya Sen (2004) has remarked, "Seeing people in terms of only their needs may give us a rather meagre view of humanity." Sen suggests that the ethical basis for value judgments about the environment can be deepened by embracing a broader range of human values. For example, people may believe that we have a responsibility to safeguard the existence of other species (as an illustration, Sen mentions the spotted owl of the Pacific northwest) regardless of whether the species in question serves any practical human needs. In other words, people may value nature for intrinsic as well as instrumental reasons. If so, environmental quality can be seen as an end in itself, and not merely a means to other ends.

Sen also observes that people hold multiple values, and that these cannot be readily reduced to a summary measure such as overall fulfillment of human needs. For example, we may believe that future generations have the right to breathe clean air, and that infringement of this right cannot be adequately compensated by improvements in other dimensions of well-being. Not everything of value can be calibrated on a single scale.

This broader ethical framework – in which "we think of human beings as agents, rather than merely as patients," in Sen's words – implies a central role for citizenship in addressing environmental challenges. Freedom to make value judgments about environmental change, and rights to a clean and safe environment, are themselves important ethical objectives. This is not only a matter of moral vision, but also of practical politics. The extent to which people are able to act as citizens depends on how power is structured and distributed in society.

In this chapter, I will use the term "environmental harm" to mean impacts on the natural environment that reduce human well-being, with the latter understood to extend beyond needs to the wider canvas of values and rights. To say that actions are bad (or good) for the environment is to say that they are bad (or good) for humankind.

This ethical stance makes no pretense of impartiality: it is unabashedly human-centered. I do not regard *Homo sapiens* as just another species, whose well-being is of no greater consequence than that of any other. In many cases, what is good for other species is good for humans, too. But not always. Sanitation and clean water-supply systems, for instance, improve human well-being by killing bacteria and other pathogens. I regard this as a good thing, an environmental improvement. The eradication of smallpox – the deliberate extinction of a virus species – counts as a good thing, too. This stance does not imply a willful disregard for nature, nor indifference to the fates of other species. On the contrary, an environmental ethics grounded in human well-being recognizes that we are a part of nature, not apart from it.

THREE QUESTIONS

Whenever we analyze economic activities that generate environmental harm, we can pose three very basic questions:

- *Who benefits* from the economic activities that cause the harm? If no one benefits – or at least thinks they do – the activities would not occur.

- *Who suffers environmental harm?* If no one is hurt by these activities, they are not a problem – at least, not in terms of human well-being.
- *Why* is the first group able to impose environmental harm on the second? That is, what allows some people to benefit at the expense of others?

The last of these questions is crucial to understanding the reasons for environmental harm. There are three possible answers to it.

One possibility is that those who are harmed belong to *future generations,* who are not here to defend themselves. In this case, the only remedy is to cultivate an ethic of inter-generational responsibility, one founded on a moral commitment to safeguard the well-being of our children and generations to come.

The second possibility is that those who are harmed lack *information.* They may know that their children are falling ill, for example, but not know what environmental circumstances are making them sick or who is responsible for them. In this case, the solution lies in greater access to information: environmental education in general and right-to-know laws in particular. In the United States, for example, the Emergency Planning and Community Right-to-Know Act of 1986, passed in the wake of the chemical plant disaster in Bhopal, India, created the Toxics Release Inventory, which makes information on releases of toxic chemicals by industrial facilities available to the public.

The final possibility is that those who are harmed are alive today and well aware of the costs imposed on them, but lack the *power* to prevail in making social decisions about the environment. In this case, the solution lies in redistributing power, so that those who suffer environmental harm are better able to defend themselves – and the environment – from others who benefit from activities that cause the problem.

PURCHASING POWER AND POLITICAL POWER

Human beings are socially differentiated in terms of wealth and influence. Differences in wealth translate into differences in *purchasing power.* Differences in influence translate into differences in *political power.*

In this respect humans are different from other species. Consider, for example, pondweed, a plant species sometimes used by ecologists to illustrate the perils of exponential growth (see, for example, Brown, 1978). Assume that the weed doubles in volume every day, and that in 30 days it will fill the pond completely, making further growth impossible and perhaps overwhelming ecological balances vital to the continued existence of the pondweed itself. When, the ecologist asks, is the pond half full? The answer, of course, is the 29th day. This parable is invoked to depict human pressure on the carrying capacity of planet Earth: metaphorically speaking, we are nearing the end of the month.

Each pondweed organism is pretty much like any other. But humans differ greatly from one another, both in their impacts on the environment and in their ability to shield themselves from these impacts. The pondweed analogy deflects attention from these differences, and from how they affect our interactions with nature.

To understand how inequalities among humans contribute to environmental harm, we need to look more closely at the two types of power: purchasing power and political power.

Purchasing Power

In a market economy, people vote on what to produce in proportion to the money they spend. Economists call this "effective demand." This differs from simple desire or need. A person can be hungry, and in that sense have demand for food, but she does not have *effective* demand – the ability to vote in the marketplace – unless hunger in her stomach is backed up by money in her pocket. The distribution of purchasing power determines how much of society's resources will be devoted to producing rice and beans, and how much to producing champagne and luxury automobiles.

Purchasing power plays a central role in describing what happens in markets. In cost–benefit analysis, it also plays a central role in *prescribing what should happen* if and when the government intervenes to correct "market failures" that arise in cases of public goods, like highways, schools, and national defense, and in cases of externalities, like pollution, that affect people who are not party to the market exchange between buyer and seller.

When the government promulgates regulations to curtail pollution, for example, it must confront the practical question: how much pollution is too much? It would be nice to live in a world with no pollution whatsoever, but, as economists are quick to point out, cutting pollution has costs as well as benefits. People need and want to eat, wear clothes, use medicines, move about, and so on, and producing these goods and services often produces some pollution, too. Faced with the choice between more goods and services and less pollution, and informed by the principle of diminishing returns (the more we have of anything, the less each additional increment is worth), the economist tells the government to aim for the "optimal level of pollution," defined as the point at which society's benefit from additional pollution reduction equals its cost in terms of foregone consumption of other goods and services.

The phrase "optimal level of pollution" rankles many environmentalists, but it is hard to argue with the logical proposition that the costs and benefits of any course of action ought to be weighed against each other. The problem is how to *measure* all the relevant costs and benefits. The economist has a toolkit for this purpose: cost–benefit analysis. It translates all costs and benefits into a single unit of measurement, money. Economists have devised ingenious ways to translate non-market values – such as the value of cleaner air or the existence of the spotted owl – into monetary terms. Contingent valuation surveys, for example, are used to ask people how much they would be willing to spend for environmental quality – say, to protect an endangered species. Hedonic regression analysis, another popular technique, uses actual market data to infer implicit prices; for example, by analyzing how housing prices vary with distance from an airport (controlling for other variables like the size of the house), in order to measure the cost of noise pollution.

The foundation for the valuation techniques of cost–benefit analysis is *willingness to pay*. The costs of environmental harms are measured by how much people are willing to pay to avoid them. This is how demand for goods and services is measured in the marketplace, so there is a certain consistency in using the same criterion to measure demand for environmental quality, and to use the results in making public policy. The willingness-to-pay criterion for valuation means, however, that the needs and desires of some people count more than the needs and desires of others – not necessarily because their desire for clean air or water is any stronger, but because they wield more purchasing power to back up their preferences.

Behind differences in willingness to pay lie differences in ability to pay. If my willingness to pay for gold mined near your community is high, and your community's ability (and hence willingness) to pay to protect its air and water from pollution by mining operations is low, then by the logic of the cost–benefit analyst, I should get the gold and you should get the pollution. In this way, differences in purchasing power can affect not only decisions made by private parties in response to market signals, but also public-policy decisions made by governments.

Political Power

In practice, real-world political systems do not faithfully adhere to the prescriptions of cost–benefit analysts. Individuals, groups, and classes differ from each other not only in their purchasing power but also in their political power. The latter includes differences in their ability to influence social decisions on environmental policies. As a result, some costs and benefits may count more than others.

Political power takes various forms:

- *decision power* to prevail in contests to determine what decision makers, both public and private, will or will not do;
- *agenda power* to keep questions off (or on) the table of the decision makers;
- *value power* to shape others' preferences to coincide with one's own; and
- *event power* to alter the circumstances that others face – for example, by blowing smoke into the atmosphere – thus presenting them with a *fait accompli.*

Each of these forms of power can lead to decisions that diverge from the optimum prescribed by cost–benefit analysis (for discussion, see Bartlett, 1989; Boyce, 2002).

If political power were distributed equally across the population, and social decisions were based simply on cost–benefit calculations, then purchasing power would be the only dimension of human differentiation that matters for environmental decisions. Once we recognize, however, that political power in practice is unequally distributed, and that it tends to be correlated with purchasing power – that is, wealth and political influence generally go together – then both dimensions of social differentiation matter, and reinforce each other.

THE ENVIRONMENTAL IMPACT OF INEQUALITY: TWO HYPOTHESES

Two hypotheses can be advanced about the environmental impact of inequalities in the distribution of purchasing power and political power:

- First, environmental harm is not randomly distributed across the population, but instead reflects the distribution of wealth and power.

The relatively wealthy and powerful tend to benefit disproportionately from the economic activities that generate environmental harm. The relatively poor and powerless tend to bear a disproportionate share of the environmental costs.

- Second, the total magnitude of environmental harm depends on the extent of inequality. Societies with wider inequalities of wealth and power will tend to have more environmental harm. Conversely, societies with relatively modest degrees of economic and political disparities will tend to have less environmental harm.

Environmental Injustice

The first hypothesis operates on both the benefit side and the cost side of the coin. Benefits from economic activities that inflict environmental harm accrue to consumers insofar as the savings from cost externalization (for example, releasing toxic chemicals out the smokestack rather than spending money on pollution control) are passed to them in the form of lower prices. Benefits accrue to the owners of firms insofar as they are able to capture these savings in the form of higher profits. On the consumer side, the rich generally get a bigger share of the benefits, for the simple reason that they consume more than the poor. On the producer side, again they get a bigger share of the benefits, since they own more productive assets, including corporate stocks. For these reasons, no matter what the division of gains between consumers and firms, the rich reap the largest share. Even if the costs of environmental harm were equally shared by all – for example, if everyone breathes the same polluted air and drinks the same polluted water – this would skew the *net* benefits from environmentally harmful economic activities in favor of the wealthy.

In practice, many environmental costs are localized, rather than being uniformly distributed across space. This makes it possible for those who are relatively wealthy and powerful to distance themselves from environmental harm caused by economic activities (Princen, 1997). Within a metropolitan area, for example, the wealthy can afford to live in neighborhoods with cleaner air and more environmental amenities. Furthermore, sometimes there are private substitutes for public environmental quality. In urban India, for instance, where public water supplies are often contaminated, the upper and middle classes can afford to consume bottled water. The poor cannot. In such cases, because access to private substitutes is based on ability to pay, again the rich are better able to avoid environmental harm.

A substantial literature on environmental justice in the United States has documented the fact that low-income people and communities of color (that is, communities with above-average percentages of non-white and non-Anglo residents) often bear disproportionate environmental harms – see, for example, the chapters in this volume by Bullard (2007), Mohai (2007), and Taylor (2007); see also Szasz and Meuser (1997), Bullard and Johnson (2000), Pastor (2003), and Boyce (2007). These findings are consistent with the first hypothesis. A number of these studies have found that race and ethnicity matter, even when controlling for income: communities with higher percentages of African-Americans, Latinos, Asian-Americans, and Native Americans tend to face greater environmental hazards – see, for example, Bouwes, Hassur, and Shapiro (2003) and Ash and Fetter (2004). This finding suggests that political power (which is correlated with race and ethnicity in the United States) has an impact on exposure to environmental harm, above and beyond whatever can be explained simply by differences in purchasing power.

Even in cases of environmental harm from which there is no escape – widely dispersed pollutants and global climate change are examples – those who are relatively poor and powerless tend to be most vulnerable. Living closest to the margin of survival, they have least ability to withstand adversity. They have less ability to afford remedial measures, like health care. And they have less political clout to secure remedial actions from government authorities. Similar vulnerability disparities are revealed by natural disasters, as when Hurricane Katrina hit New Orleans in August 2005. "In a sense, environmental justice is about slow-motion disasters," in the words of a recent study of Katrina's impact, "and disasters reveal environmental injustice in a fast-forward mode" (Pastor et al., 2006, p. 9).

More Inequality, More Harm?

The second hypothesis – that more inequality causes more environmental harm overall – may be less intuitively evident than the first. Inequalities of wealth and power could have two opposing effects. When the beneficiaries from environmentally harmful activities are more powerful than those who bear their costs, greater inequality can be expected to result in more environmental harm. On the other hand, when those who bear the costs are more powerful than the beneficiaries, we might expect the opposite: greater inequality yields less environmental harm.

Which scenario is more common? The second one certainly occurs, for example, when African tribespeople are expelled from their traditional hunting grounds on the grounds that their activities are environmentally harmful, in order to create protected areas for the enjoyment of affluent foreign tourists, a phenomenon that has been labeled "coercive conservation" (see, for example, Peluso, 1993; Neumann, 2001; Mulder & Coppolillo, 2005, pp. 31–37). The slum clearance program in Bangladesh described in the prologue is another example. But there are good reasons to believe that the first scenario is far more prevalent. If, as I have argued above, the benefits of environmentally harmful activities flow disproportionately to the relatively well-off by virtue of their higher consumption and capital ownership, and purchasing power is correlated with political power, it follows that the beneficiaries of these activities tend to be more powerful than those who bear net costs – in which case, wider inequalities can be expected to translate into greater environmental harm.

In a statistical test of the second hypothesis, Boyce, Klemer, Templet, and Willis (1999) found that among the 50 U.S. states, those with more equitable distributions of power (measured by voter participation, educational attainments, tax fairness, and Medicaid access) tend to have stronger environmental policies and better environmental outcomes. Further evidence in support of this hypothesis comes from a study of the relationship between residential segregation and cancer risks from air pollution in the United States, which found that greater segregation on racial and ethnic lines is correlated with worse environmental and health outcomes for all groups, not only for people of color (Morello-Frosch & Jesdale, 2006).

Similarly, cross-country studies at the international level have found that a more equitable distribution of power – measured by such variables as democracy, political and civil rights, and adult literacy – is correlated with better environmental quality, even while controlling for other variables such as differences in per capita income (see, for example, Torras & Boyce, 1998; Barrett & Graddy, 2000; for a review of these and other studies, see also Boyce, 2007).

In sum, both theoretical reasoning and empirical evidence support the conclusion that inequality is bad for the environment. People are not like pondweed. How we treat the natural environment depends on how we treat each other.

ROOM FOR HOPE

There is another important way that humans differ from pondweed: we have brains. Indeed we are exceptional among all species in our ability to accumulate knowledge, pass it from one generation to the next, and change our behavior accordingly. This includes knowledge about our interactions with the natural environment and with each other.

Respecting Nature's Limits

We can learn how to respect nature's limits, and thus how to limit environmental harm – if we choose to do so. We can learn about the growth rates of renewable natural resources, such as trees in forest and fish in the sea, and we can manage our own harvests of these resources to ensure sustainable yields. We can learn about nature's finite stocks of non-renewable resources, such as minerals and fossil fuels, and we can develop recycling and renewable alternatives to avert future shortages. We can learn about the limited capacity of air, lands, and water bodies to safely absorb and break down wastes, and we can limit the rates at which we discharge pollutants accordingly.

As an illustration, consider our response to the threat posed by chemicals that were depleting the Earth's protective ozone layer, exposing life on the planet to increasing levels of ultraviolet-B radiation. The danger was not recognized until the early 1970s, when scientists first hypothesized that chlorofluorocarbons (CFCs), man-made compounds used for a variety of purposes, including aerosol propellants and air conditioner coolants, were breaking down ozone molecules in the Earth's stratosphere. The harm was invisible but insidious. In a remarkable instance of international co-operation, by 1987 the nations of the world had agreed to curtail their CFC emissions via the Montreal Protocol (Haas, 1992). No other species is capable of such conscious self-regulating behavior.

Of course, to say that we can modify our actions on the basis of knowledge about nature's limits does not mean that we necessarily will do so. But if we choose to act, we can. And, as the Montreal Protocol illustrates, sometimes we do. The question is, why do we act to protect the environment in some times and places, and not in others? The answer, I believe, is that whether and how we act (or fail to act) depends on the balance of power in the present generation between those who benefit by ignoring nature's limits and those who pay the price, and on whether we embrace an ethic of responsibility toward future generations.

Nor do I wish to imply that humans are omniscient, understanding fully the consequences of our actions. Had scientists been a few decades slower to grasp the environmental implications of CFC emissions, we might not have recognized the threat until it was too late. We need to understand not only nature's limits, but also the limits of our own knowledge. Given the uncertainties and unknowns about the environmental impacts of our actions, prudence demands that we adopt a "precautionary" approach to environmental policy (for discussions, see Harremoës et al., 2002; Dorman, 2005).

Investing in Nature's Wealth

Humans can not only deplete nature's wealth; we also can increase it. If our value system is founded on long-term human well-being – if this is the basis on which we compare states of the world, and define what is good and bad for the environment – then we can improve the environment as well as harm it.

There are three ways that humans invest in nature's wealth:

- *Ecological restoration* repairs past harms. Examples include the reforestation of deforested landscapes; the replenishment of depleted fisheries; the clean-up of contaminated soils and water bodies; and the restoration of degraded wildlife habitat.[2]
- *Co-evolution* refers to human modifications that create an environment that is better able to support long-term human well-being. One example is "soil banking": farming practices that build deeper and more fertile soils, such as *terra preta do indio* ('dark earth of the Indians') in Amazonia (Mann, 2002) and those of the *acequia* landscape mosaic in the upper Rio Grande bioregion of the southwestern United States (Peña, 2003; for more examples of soil banking, see Brookfield, 2001, pp. 96–97). Another example – arguably the most valuable investment in nature's wealth in human history – is the domestication of plants and animals that began some 10,000 years ago, and the subsequent evolution of genetic diversity in crops and livestock (see Boyce, 2006).
- *Environmental preventive health* refers to measures to reduce the prevalence of pathogens and disease-bearing insects. One example,

already mentioned, is the eradication of the smallpox virus through an international effort that culminated in the mid-1970s. Another example is the modification of aquatic habitats to reduce mosquito populations, a form of investment that played a major role in eliminating malaria from Europe and North America (Kitron & Spielman, 1989; Willott, 2004).

In all three ways, human beings can and sometimes do improve the environment, from the standpoint of long-term human well-being. Humans are not necessarily a blight on the face of the planet, a cancer that ultimately will destroy its host. We have learned a great deal about how to respect nature's limits and invest in nature's wealth, and we have the capacity to learn more. In our dealings with nature, there is room for hope.

Making Social Change

The inequalities of power and wealth that generate environmental harm are not forces of nature. Political and economic disparities are social constructions, and as such they can be reconstructed.

To be sure, there is no certainty that social change will proceed inexorably toward more democratic distributions of political power and more egalitarian distributions of purchasing power. It is all too easy to find past and present examples of movements in the opposite direction. But to say that something is not inevitable is not to say that it is impossible.

In fact, an even stronger claim is possible: there is ample evidence that the overall trend in human history, notwithstanding periodic reversals, is toward more equality in our social arrangements. Only three centuries ago, monarchs and aristocracies ruled most of the world. A century and a half ago, slavery was still legal in much of the United States. The state of Massachusetts, the first in the country to mandate free primary education for all children, did so only in 1852; it was not until 1918 that all states had followed suit. The amendment to the U.S. constitution that granted women the right to vote was adopted less than a century ago. In much of Asia and Africa, colonial rule ended only two generations ago. It has been little more than a decade since apartheid ended in South Africa. "The arc of the moral universe is long," Dr. Martin Luther King remarked in his 1965 commencement address at Oberlin College, "but it bends toward justice" (King, 1965).

There is room for hope in our dealings with each other, too.

ONE WORLD, READY OR NOT

Both the prospects and need for changes in our relationships with nature and each other are affected by changes in the scale at which human interactions occur. For much of human history, the implications of the fact that we share a single planet were hidden from view by spatial fragmentation. This slowly changed over time, especially with the development of agriculture, states, and more effective means of transport beginning some ten millennia ago. The pace of change accelerated in the past few centuries, in the process nowadays dubbed "globalization." Today the fact that we live in one world is not only a physical reality, but also an economic, cultural, and political reality.

Uneven Globalization: Markets and Governance

Globalization has proceeded most rapidly in the economic arena. Around the world, production and consumption are being increasingly integrated into a single market. Indeed, for many the term "globalization" has come to signify not only the process of economic integration, but also the subordination of more and more economic activity across the globe to the laws of the market. The extension of the market brings tangible benefits, as Adam Smith famously observed in *The Wealth of Nations*: responding to price signals, decentralized producers are guided by an "invisible hand" to specialize in what they can make most cheaply, unleashing impressive productivity gains.

But the market also has important limitations:

- *Market failure:* One way to make goods cheaply is to push costs onto others – generating what economists call "negative externalities." For example, firms that do not spend money on pollution control may enjoy a competitive advantage over firms that do. As markets extend their reach, the costs of such market failures can grow alongside the benefits of specialization.
- *Fairness:* The production and distribution of goods and services for the market are driven by effective demand – that is, willingness to pay backed by ability to pay. An inequitable distribution of purchasing power leads to an inequitable distribution of resources.
- *Resilience:* The market pursues a logic of short-term optimization: lowest-cost producers using the "best" technology can undersell rivals and

ultimately drive them out of business. As a result, producers tend to converge on the same technology. Yet resilience – the ability to withstand shocks and adapt to changing circumstances – requires a range of alternative technological options (Rammel & van den Bergh, 2003).

- *Moral capital:* Finally, by elevating a narrowly conceived "self-interest" above all other values, markets may lead to depreciation of moral capital that is crucial to the functioning of society. In fact, widespread commitment to moral precepts is necessary for markets themselves to function, since respect for rights and contractual obligations typically rests not on self-interest but instead on accepted norms about what is the right thing to do (for discussion, see Basu, 1983; Sen, 1986; Bell, 1996).

For all four reasons, the globalization of markets needs to be complemented by the globalization of governance. The latter includes not only formal international institutions and inter-governmental agreements but also informal governance by non-state institutions and networks. Many of the problems associated with globalization arise from the fact that it has been uneven: the development of global governance has lagged behind the development of global markets (Young, 1994; Boyce, 2004).

NAFTA and the Environment: A Case Study in Uneven Globalization

The environmental consequences of uneven globalization can be illustrated by looking at the effects of the North American Free Trade Agreement (NAFTA), the free trade agreement among the United States, Mexico, and Canada that went into effect in 1994. In the early 1990s, the debate over NAFTA split the U.S. environmental movement. Some maintained that the trade agreement would promote "harmonization upwards," a continental convergence to higher environmental standards, by generating higher incomes and stronger demand for environmental protection in Mexico. Others contended that it would spark a "race to the bottom," as firms moved (or threatened to move) south of the U.S.–Mexico border to take advantage of lax environmental regulations.

Both sides in the debate shared one premise: environmental practices in Mexico were evidently inferior to those in the United States and Canada. This assumption helps to explain why few environmentalists voiced concern about what in the end may turn out to be NAFTA's most profound environmental impact: the erosion of Mexico's rich heritage of genetically diverse maize varieties by cheap corn imported from the United States.

Maize ('corn' in U.S. parlance) is the single most important crop in both countries.[3] On the eve of NAFTA, U.S. maize was sold at roughly $110 per ton at the border, whereas Mexican growers were receiving $240 per ton for their crops. With the dismantling of trade barriers, the Mexican price is now converging to the lower U.S. price, hitting hard at the livelihoods of Mexican *campesinos.*[4]

By the measuring stick of market prices, U.S. farmers are more efficient than Mexican growers. But this competitive edge results, in no small measure, from the neglect of market failures on both sides of the border, as well as from government subsidies and natural advantages such as more regular rainfall in the U.S. corn belt (Boyce, 1996). U.S. corn production relies on massive applications of pesticides, fertilizers, and energy inputs, all of which generate substantial environmental harm. The resulting costs do not figure in the market price.

At the same time, the *campesino* farmers of southern and central Mexico today provide a great "positive externality" to humankind by sustaining genetic diversity in one of the world's most important food crops. In their small plots, where corn was first domesticated some seven millennia ago, the maize plant continues to evolve via the process Darwin called "artificial selection," as farmers select seeds for the following year's crop from the plants that perform best in the face of changing conditions. Mexican farmers still grow thousands of varieties of maize. In the United States, by contrast, fewer than a dozen varieties now account for half of the country's total corn acreage.

As a result of its low diversity, the U.S. corn crop has a high degree of genetic vulnerability – the eggs-in-one-basket syndrome – a problem dramatically revealed in 1970 when a new strain of leaf blight destroyed one-fifth of the nation's harvest. In the effort to stay ahead of the insects and plant diseases that evolve rapidly in genetically uniform fields, U.S. plant breeders engage in a "varietal relay race," constantly seeking to develop new resistant varieties. The average commercial lifespan of a corn variety in the United States is only seven years, after which it is replaced by new ones. In effect, modern agriculture substitutes diversity through time for diversity at any point in time. The raw material that plant breeders use in this relay race is the genetic diversity that has been bequeathed to us by generations of small farmers in Mexico and elsewhere.[5]

The irony is that under NAFTA, the success of U.S. corn production in the marketplace is undermining the genetic base on which its own long-term viability depends. The globalization of the market is being accompanied by the globalization of market failure, and a loss of resilience. In the short term, the main people harmed by this process are the Mexican *campesinos*, who lack the political power to ensure that their investments in nature's wealth are rewarded. In the long term, those harmed are future generations around the world, whose food security is being undermined by the erosion of crop genetic diversity.

Remedying this environmental harm will require more than defensive actions by single individuals, communities, or states, important though these may be. Ultimately this and other transnational environmental problems require building institutions that bridge the gap between the globalization of markets and the globalization of governance (for discussion of formal and informal institutions that could help to conserve crop genetic diversity, see Mann, 2004; Boyce, 2006).

CONCLUDING REMARKS

In this chapter, I have explored some of the implications of recognizing that environmental harm is not a random by-product of economic activities, but instead a cost that is imposed on some and that benefits others. Rectifying the market failures and governance failures that lead to environmental harm requires repairing the disparities of wealth and power that enable these failures.

This does not mean that greater equality is a panacea for all environmental ills. A more democratic distribution of power and a more egalitarian distribution of wealth are not all that is needed to prevent environmental harm. To say that these are necessary does imply that they are sufficient. Safeguarding the natural environment also will require us to cultivate an ethic of moral responsibility to others – particularly when the costs of environmental harm would primarily fall upon future generations.

We can have a healthy environment, and bequeath one to future generations, by respecting nature's limits and investing in nature's wealth. Achieving these goals does not only require rebalancing our relationships with nature. It will also require rebalancing our relationships with our fellow humans.

NOTES

1. For accounts of the famine and its causes, see Sen (1981) and Ravallion (1987). For an account of the life in the village, see Hartmann and Boyce (1979, 1983).

2. For examples, see Narain and Agarwal (2007) on water harvesting in semi-arid zones of rural India, and Rahman and Minkin (2007) on the rehabilitation of inland fisheries in Bangladesh. A distinction is sometimes made between restoration and rehabilitation, with the former referring to returning an ecosystem to "its historic trajectory" and the latter to "the reparation of ecosystem processes, productivity and services" (Society for Ecological Restoration, 2004). If humans are regarded as an alien species – if we are truly apart from nature, not a part of it – this distinction makes sense: one may imagine the "historic trajectory" of ecosystems in the absence of any human impacts whatsoever. In my view, this is a peculiar view of history. In any event, proponents of this distinction have concluded that restoration "probably encompasses a large majority of project work that has previously been identified as rehabilitation."

3. Michael Pollan (2006, pp. 22–23) observes that once the cycling of corn through animals is counted, the average American today has more corn in his diet than the average Mexican.

4. For further discussion, see Boyce (1996, 2004). So far, the sharp decline in Mexican corn production that was predicted by many as a result of NAFTA has not occurred, apparently because economic opportunities elsewhere in the Mexican economy have been so scarce (see Ackerman, Wise, Gallagher, Ney, & Flores, 2003).

5. Samples of many Mexican maize varieties are stored in "seed banks" at agricultural research institutes. But seed banks are insecure, being subject to the perennial hazards of inadequate funding, accidents, and war. Moreover, having seeds in the bank is not the same as knowing about varietal properties such as pest resistance and climate sensitivity, information that is most readily obtained in the field. And even at best, seed banks can conserve only the existing stock of genetic diversity; they cannot replicate the ongoing process of evolution that takes place in the farmers' fields. For further discussion of the value of in situ (in-the-field) crop genetic diversity, see Brush (2003) and Boyce (2006).

REFERENCES

Ackerman, F., Wise, T. A., Gallagher, K. P., Ney, L., & Flores, R. (2003). Free trade, corn, and the environment: Environmental impacts of US–Mexico corn trade under NAFTA. Working Paper No. 03-06. Tufts University, Global Development and Environment Institute, Medford, Massachusetts.

Ash, M., & Fetter, T. R. (2004). Who lives on the wrong side of the environmental tracks? *Social Science Quarterly, 85*(2), 441–462.

Barrett, S., & Graddy, K. (2000). Freedom, growth, and the environment. *Environment and Development Economics, 5,* 433–456.

Bartlett, R. (1989). *Economics and power: An inquiry into human relations and markets.* Cambridge: Cambridge University Press.

Basu, K. (1983). On why we do not try to walk off without paying after a taxi ride. *Economic and Political Weekly, 18,* 2011–2012.

Bell, D. (1996). *The cultural contradictions of capitalism (2nd ed.).* New York: Basic Books.

Bouwes, N., Hassur, S., & Shapiro, M. (2003). Information for empowerment: The EPA's risk-screening environmental indicators project. In: J. K. Boyce & B. G. Shelley (Eds), *Natural assets: Democratizing environmental ownership*. Washington, DC: Island Press, ch. 6.

Boyce, J. K. (1996). Ecological distribution, agricultural trade liberalization, and in situ genetic diversity. *Journal of Income Distribution*, 6(2), 263–284.

Boyce, J. K. (2002). *The political economy of the environment*. Northampton: Edward Elgar.

Boyce, J. K. (2004). Green and brown? Globalization and the environment. *Oxford Review of Economic Policy*, 20(1), 105–128.

Boyce, J. K. (2006). A future for small farms? Biodiversity and sustainable agriculture. In: J. K. Boyce, S. Cullenberg, P. K. Pattanaik & R. Pollin (Eds), *Human development in the era of globalization: Essays in honor of Keith B. Griffin* (pp. 83–104). Northampton: Edward Elgar.

Boyce, J. K. (2007). Inequality and environmental protection. In: J.-M. Baland, P. Bardhan & S. Bowles (Eds), *Inequality, collective action, and environmental sustainability* (pp. 314–348). Princeton: Princeton University Press.

Boyce, J. K., Klemer, A. R., Templet, P. H., & Willis, C. E. (1999). Power distribution, the environment, and public health: A state-level analysis. *Ecological Economics*, 29, 127–140. Reprinted in Boyce (2002, ch. 6).

Brookfield, H. (2001). *Exploring agrodiversity*. New York: Columbia University Press.

Brown, L. (1978). *The twenty-ninth day: Accommodating human needs and numbers to the Earth's resources*. New York: Norton.

Brush, S. B. (2003). The lighthouse and the potato: Internalizing the value of crop genetic diversity. In: J. K. Boyce & B. G. Shelley (Eds), *Natural assets: Democratizing environmental ownership*. Washington, DC: Island Press, ch. 10.

Bullard, R. D., & Johnson, G. S. (2000). Environmental justice: Grassroots activism and its impact on public policy decision making. *Journal of Social Issues*, 56(3), 555–578.

Dorman, P. (2005). Evolving knowledge and the precautionary principle. *Ecological Economics*, 53, 169–176.

Haas, P. (1992). Banning chlorofluorocarbons: Epistemic community efforts to protect stratospheric ozone. *International Organization*, 46(1), 187–224.

Harremoës, P., et al. (2002). *The precautionary principle in the 20th century: Late lessons from early warnings*. London: Earthscan, for the European Environment Agency.

Hartmann, B., & Boyce, J. K. (1979). *Needless hunger: Voices from a Bangladesh village*. San Francisco: Institute for Food and Development Policy.

Hartmann, B., & Boyce, J. K. (1983). *A quiet violence: View from a Bangladesh village*. London/ San Francisco: Zed Press/Institute for Food and Development Policy.

King, M. L. (1965). Staying awake through a great revolution. Commencement address at Oberlin College, Oberlin, Ohio, June.

Kitron, U., & Spielman, A. (1989). Suppression of transmission of malaria through source reduction: Antianopheline measures applied in Israel, the United States, and Italy. *Reviews of Infectious Diseases*, 11(3), 391–406.

Mann, C. (2002). The real dirt on rainforest fertility. *Science*, 297, 920–923.

Mann, C. (2004). *Diversity on the farm*. New York/Amherst, Massachusetts: Ford Foundation/ Political Economy Research Institute. Available at http://www.peri.umass.edu/ fileadmin/pdf/Mann.pdf

Morello-Frosch, R. A., & Jesdale, B. (2006). Separate and unequal: Residential segregation and estimated cancer risks associated with ambient air toxics in US metropolitan areas. *Environmental Health Perspectives, 114*(3), 386–393.

Mulder, M. B., & Coppolillo, P. (2005). *Conservation: Linking ecology, economics, and culture.* Princeton: Princeton University Press.

Narain, S., & Agarwal, A. (2007). Harvesting the rain: Fighting ecological poverty through participatory democracy. In: J. K. Boyce, S. Narain & E. A. Stanton (Eds), *Reclaiming nature: Environmental justice and ecological restoration.* London: Anthem Press, ch. 3.

Neumann, R. P. (2001). Disciplining peasants in Tanzania: From state violence to self-surveillance in wildlife conservation. In: M. Watts & N. Peluso (Eds), *Violent environments.* Ithaca: Cornell University Press, ch. 13.

Pastor, M. (2003). Building social capital to protect natural capital: The quest for environmental justice. In: J. K. Boyce & B. G. Shelley (Eds), *Natural assets: Democratizing environmental ownership.* Washington, DC: Island Press, ch. 4.

Pastor, M., Bullard, R. D., Boyce, J. K., Fothergill, A., Morello-Frosch, R., & Wright, B. (2006). *In the wake of the storm: Environment, disaster, and race after Katrina.* New York: Russell Sage Foundation.

Peluso, N. (1993). Coercing conservation: The politics of state resource control. *Global Environmental Change, 3*(2), 199–217.

Peña, D. (2003). The watershed commonwealth of the upper Rio Grande. In: J. K. Boyce & B. G. Shelley (Eds), *Natural assets: Democratizing environmental ownership.* Washington, DC: Island Press, ch. 9.

Pollan, M. (2006). *Omnivore's dilemma: A natural history of four meals.* New York: Penguin Press.

Princen, T. (1997). The shading and distancing of commerce: When internalization is not enough. *Ecological Economics, 20,* 235–253. Reprinted In: T. Princen, M. Mainates & K. Conca (Eds), *Confronting consumption,* Cambridge, MA: MIT Press, 2002, ch. 5.

Rahman, M., & Minkin, S. F. (2007). Net benefits: The ecological restoration of inland fisheries in Bangladesh. In: J. K. Boyce, S. Narain & E. A. Stanton (Eds), *Reclaiming nature: Environmental justice and ecological restoration.* London: Anthem Press, ch. 4.

Rammel, C., & van den Bergh, J. C. J. M. (2003). Evolutionary policies for sustainable development: Adaptive flexibility and risk minimising. *Ecological Economics, 47,* 121–133.

Ravallion, M. (1987). *Markets and Famines.* Oxford: Clarendon.

Sen, A. (1981). *Poverty and famines: An essay on entitlements and deprivation.* Oxford: Clarendon.

Sen, A. (1986). Adam Smith's prudence. In: S. Lall & F. Stewart (Eds), *Theory and reality in development.* London: Macmillan.

Sen, A. (2004). Why we should preserve the spotted owl. *London Review of Books, 26*(3), February 5.

Society for Ecological Restoration. (2004). *The SER International primer on ecological restoration* (Available at *www.ser.org*). Tucson: Society for Ecological Restoration International.

Szasz, A., & Meuser, M. (1997). Environmental inequalities: Literature review and proposals for new directions in research and theory. *Current Sociology, 45*(3), 99–120.

Torras, M., & Boyce, J. K. (1998). Income, inequality, and pollution: A reassessment of the environmental Kuznets curve. *Ecological Economics, 25*, 147–160. Reprinted in Boyce (2002, ch. 5).

Willott, E. (2004). Restoring nature, without mosquitoes? *Restoration Ecology, 12*(2), 147–153.

World Commission on Environment and Development. (1987). *Our common future* (the Brundtland Report). New York: Oxford University Press.

Young, O. (1994). *International governance: Protecting the environment in a stateless society.* Ithaca: Cornell University Press.

SET UP A CONTINUATION ORDER TODAY!

Did you know that you can set up a continuation order on all Elsevier-JAI series and have each new volume sent directly to you upon publication? For details on how to set up a **continuation order**, contact your nearest regional sales office listed below.

To view related series in Sociology, please visit:

www.elsevier.com/sociology

The Americas
Customer Service Department
11830 Westline Industrial Drive
St. Louis, MO 63146
USA
US customers:
Tel: +1 800 545 2522 (Toll-free number)
Fax: +1 800 535 9935
For Customers outside US:
Tel: +1 800 460 3110 (Toll-free number).
Fax: +1 314 453 7095
usbkinfo@elsevier.com

Europe, Middle East & Africa
Customer Service Department
Linacre House
Jordan Hill
Oxford OX2 8DP
UK
Tel: +44 (0) 1865 474140
Fax: +44 (0) 1865 474141
eurobkinfo@elsevier.com

Japan
Customer Service Department
2F Higashi Azabu, 1 Chome Bldg
1-9-15 Higashi Azabu, Minato-ku
Tokyo 106-0044
Japan
Tel: +81 3 3589 6370
Fax: +81 3 3589 6371
books@elsevierjapan.com

APAC
Customer Service Department
3 Killiney Road #08-01
Winsland House I
Singapore 239519
Tel: +65 6349 0222
Fax: +65 6733 1510
asiainfo@elsevier.com

Australia & New Zealand
Customer Service Department
30-52 Smidmore Street
Marrickville, New South Wales 2204
Australia
Tel: +61 (02) 9517 8999
Fax: +61 (02) 9517 2249
service@elsevier.com.au

30% Discount for Authors on All Books!

A 30% discount is available to Elsevier book and journal contributors on all books *(except multi-volume reference works)*.

To claim your discount, full payment is required with your order, which must be sent directly to the publisher at the nearest regional sales office above.